Advanced Algebra

Rings and Fields

Alan Parks
Department of Mathematics
Lawrence University

♠ The first version of this text was completed on December 27, 1987. Revised January 1990, December 1993, November 2001, June-July, December 2004, December 2006, January 2007, July 2012. Thanks to Phaik Wei Loh, who helped translate the manuscript into TeX in 1990. The author translated a revised version into LaTeX, using his own page layout macros.

♡ For Jean

Contents

Introduction.

The beginning of wisdom is this: get wisdom, and whatever you get, get insight. (Proverbs 4.7)

Eighteenth and nineteenth century mathematics distinguished itself by problem solving through abstraction: finding the right abstract context in which a variety of apparent peculiarities become special cases of a general situation that could be studied profitably for its own sake. The objects of study in this course, rings and fields, are examples of such abstract context, and the development of their properties lead to the unraveling of problems whose solutions had eluded mathematicians for centuries – problems of polynomial equations, integer number theory, and geometry. The self-evident importance of these problems should demonstrate the worth of the algebra while at the same time giving us a sense of accomplishment.

This course continues courses introducing number theory and group theory. Many of the specific facts needed are listed at the beginning of Chapter 1. Our overall purpose is to teach mathematical thought. We have chosen particular topics because they illustrate methodology: adaptation of an idea to a different context, distillation of a set of ideas, and, most important, development of intuition through examples.

In class we will discuss the proofs presented here, observing the imperative that you be convinced of the validity of each proposition and theorem. We will also spend a great deal of class time working examples. You will receive the usual problem sets, often working out more examples, and occasionally providing a proof of a result necessary to our progress in the text.

This text contains a bibliography and an extensive index. The reader will find many other books covering this material in libraries and online.

CHAPTER 1

Preliminaries

1. The Complex Numbers

We will assume that the reader is familiar with the arithmetic of finite sets. The number of elements of a finite set S is its *order*, denoted $|S|$. We have to be careful with the vertical bar notation, since it is used for several different things – absolute value, for instance. Also, the word *order* is used in different contexts.

We use standard notation for the basic sets of numbers: the set $\mathbb{Z}$ of integers, the set $\mathbb{N}$ of non-negative integers,[1] the set $\mathbb{Q}$ of rational numbers, the set $\mathbb{R}$ of real numbers and the set $\mathbb{C}$ of complex numbers. Of course, all these sets are subsets of the complex numbers. Here is a short list of assumptions we will make.

Well-ordering of the integers If S is a non-empty set of integers such that S is bounded above, then S has a maximum element. If S is a non-empty set of integers that is bounded below, then S has a minimum element.

The division theorem (This fact is also called the *quotient/remainder theorem*.) If n, m are integers with $n \neq 0$, then there are unique integers q, r such that $0 \leq r < |n|$ and $m = q \cdot n + r$.

---

[1] It is common to regard the non-negative integers and the positive integers as fundamental sets for counting and indexing. Note that we are using the non-negative integers.

Integer congruence For a positive integer n, and for integers a, b, the statement "$a \equiv b \mod n$" means that $a - b$ is an integer multiple of n. We say that a is *congruent* to b mod n. We will write $a \equiv b$ when n is clear from context.

Modular integers For each positive integer n, the set $\mathbb{Z}_n$ consists of the n distinct integers up to congruence mod n. Addition and multiplication are operations on $\mathbb{Z}_n$ inherited from the integers.

Prime factorization By an *integer prime* we will mean an integer p that is not ± 1 and whose only integer factors are $\pm p$ and ± 1. The *Fundamental Theorem of Arithmetic* asserts that each integer greater than 1 is a finite product of positive primes. The mention that we will eventually use the word *prime* in contexts other than the integers; we will call special attention to those contexts.

Greatest common divisor If m, n are integers, not both 0, then there is a maximal positive integer that divides them both – their greatest common divisor (GCD). The GCD can be written $a \cdot m + b \cdot n$ for some integers a, b.

Rational fractions A rational number is a ratio a/b for integers a, b with $b \neq 0$.

Real numbers The *Completeness Property* holds: if S is a non-empty set of real numbers, bounded above, then S has a unique least upper bound. This least upper bound is called the *supremum* (or *sup*) of the set.

Roots of real numbers If n is a positive integer and if r is a non-negative real number, there is a unique non-negative real number s such that $s^n = r$. We write $s = \sqrt[n]{r}$.

Trigonometry We will assume you know the basic properties of the real-valued functions $\cos(x)$ and $\sin(x)$. In particular, if a, b are real numbers such that $a^2 + b^2 = 1$, then there is a real number θ such that $\cos(\theta) = a$ and $\sin(\theta) = b$.

Complex numbers There is a (non-real) number i such that $i^2 = -1$, and every element of the set $\mathbb{C}$ of complex numbers can be written uniquely $a+b{\cdot}i$ where a, b are real numbers. We will give two constructions of the number i.

Our next few facts could probably be assumed as well, but we will provide proofs, feeling that they will help get the course started. For a real number θ, define

$$\exp(i \cdot \theta) = \cos(\theta) + i \cdot \sin(\theta)$$

Formally, we are defining a function of θ, and so θ should be the argument rather than $i{\cdot}\theta$. We will stick with the traditional notation using the exponential function, although we will not need to assume properties of that function – we will stick to the definition just given and to properties of the sine and cosine function. To get started, compute

$$\exp(i \cdot 0) = \cos(0) + i \cdot \sin(0) = 1 + i \cdot 0 = 1$$

We use the angle addition formulas:

$$\begin{aligned}
\exp(i \cdot \theta) \cdot \exp(i \cdot \phi) &= \Big[\cos(\theta) + i\sin(\theta) \Big] \cdot \Big[\cos(\phi) + i\sin(\phi) \Big] \\
&= \cos(\theta)\cos(\phi) - \sin(\theta)\sin(\phi) \\
&\quad + i\sin(\theta)\cos(\phi) + i\cos(\theta)\sin(\phi) \\
&= \cos(\theta + \phi) + i\sin(\theta + \phi)
\end{aligned}$$

and we obtain an identity that you may well have seen before:

$$\exp(i \cdot \theta) \cdot \exp(i \cdot \phi) = \exp(i \cdot (\theta + \phi))$$

For a complex number $z = a + i{\cdot}b$, where a, b are real, define $|z| = \sqrt{a^2 + b^2}$. Notice that when z is real (when $b = 0$), this is the ordinary absolute value of z. In general, $|z|$ is called the *modulus* of z. If $z \neq 0$, then $|z|$ is a positive real number.

Next, we define the *polar form* of a complex number: $r \cdot \exp(i \cdot \theta)$ where r is a non-negative real number. We see that

$$r \cdot \exp(i \cdot \theta) = a + i \cdot b \quad \text{where} \quad a = r \cdot \cos(\theta) \text{ and } b = r \cdot \sin(\theta)$$

Then $a^2 + b^2 = r^2$, and since $r \geq 0$, this shows that r is the modulus of $a + ib$.

An arbitrary complex number can be put into polar form. Indeed, $0 = 0 \cdot \exp(i \cdot \theta)$ for all θ, and if $z = a + i \cdot b$ is not 0, then notice that

$$\left(\frac{a}{|z|}\right)^2 + \left(\frac{b}{|z|}\right)^2 = \frac{a^2 + b^2}{|z|^2} = 1$$

As we mentioned above, this identifies a real number θ such that

$$\frac{a}{|z|} = \cos(\theta) \quad \text{and} \quad \frac{b}{|z|} = \sin(\theta)$$

Recalling that $z = a + i \cdot b$, we see that

$$z = |z| \cdot \cos(\theta) + i \cdot |z| \cdot \sin(\theta) = |z| \cdot \exp(i \cdot \theta)$$

and this gives a polar form for z.

It is simple but interesting to consider the case that z is a real number. If $z \geq 0$, then

$$z = z \cdot \exp(i \cdot 0)$$

is the polar representation. If $z < 0$, then

$$z = (-z) \cdot \exp(i \cdot \pi)$$

does the trick.

You might know that the complex number $a + i \cdot b$ can be identified with the point (a, b) in the plane. A polar form $r \cdot \exp(i \cdot \theta)$ of $a + i \cdot b$ corresponds to polar coordinates (r, θ) for (a, b).

Polar representation makes it easy to multiply complex numbers.

$$r \cdot \exp(i \cdot \theta) \cdot s \cdot \exp(i \cdot \phi) = (r \cdot s) \cdot \exp(i \cdot (\theta + \phi))$$

An induction argument then proves the following, for all non-negative integers n and all real numbers r, θ:

$$\tag{1.1} [r \cdot \exp(i \cdot \theta)]^n = r^n \cdot \exp(i \cdot n \cdot \theta)$$

It follows that the same identity holds when $n < 0$, as long as $r \neq 0$. Thus, we have (1.1) for all integers n when $r \neq 0$.

For a complex number z, we will also need its *complex conjugate* denoted $\overline{z}$. For $z = a + i \cdot b$ where $a, b \in \mathbb{R}$, we define $\overline{z} = a - i \cdot b$. We collect the properties of the modulus and the conjugate we will need.

PROPOSITION 1.1. *If $z, w \in \mathbb{C}$, then $|z \cdot w| = |z| \cdot |w|$ and $\overline{z + w} = \overline{z} + \overline{w}$ and $\overline{z \cdot w} = \overline{z} \cdot \overline{w}$. If $z = |z| \cdot \exp(i \cdot \theta)$ for a real number θ, then $\overline{z} = |z| \cdot \exp(-i \cdot \theta)$. We have $z \cdot \overline{z} = |z|^2$ and we have $\overline{\overline{z}} = z$.*

PROOF. Write $z = |z| \cdot \exp(i \cdot \theta)$ and $w = |w| \cdot \exp(i \cdot \phi)$, and compute

$$z \cdot w = |z| \cdot |w| \cdot \exp(i \cdot (\theta + \phi))$$

and since $|z| \cdot |w|$ is non-negative, we see that $|z \cdot w| = |z \cdot w| = |z| \cdot |w|$.

Recall[2] that $\cos(-\theta) = \cos(\theta)$ and $\sin(-\theta) = -\sin(\theta)$. It follows directly that

$$|z| \cdot \exp(-i \cdot \theta) = |z| \cdot \cos(-\theta) + |z| \cdot \sin(-\theta)$$
$$= |z| \cdot \cos(\theta) - |z| \cdot \sin(\theta)$$

so that the conjugate of $|z| \cdot \exp(i \cdot \theta)$ is $|z| \cdot \exp(-i \cdot \theta)$. In light of this,

$$\overline{z} \cdot \overline{w} = |z| \cdot \exp(-i \cdot \theta) \cdot |w| \cdot \exp(-i \cdot \phi)$$
$$= |z \cdot w| \cdot \exp(-i \cdot (\theta + \phi))$$
$$= \overline{z \cdot w}$$

---

[2]The two identities in this sentence follow from the angle addition formulas: for instance, write $\cos(-\theta) = \cos(0 - \theta)$.

We also see that

$$z \cdot \bar{z} = |z| \cdot \exp(i \cdot \theta) \cdot |z| \cdot \exp(-i \cdot \theta)$$
$$= |z|^2 \cdot \exp(i \cdot \theta - i \cdot \theta) = |z|^2$$

The formula $\overline{z + w} = \bar{z} + \bar{w}$ follows directly from the definition of the conjugate, as does the fact that $\bar{\bar{z}} = z$. $\square$

Finally, we will prove an important theorem about solving for n-th roots of complex numbers. Since the only way to have $x^n = 0$ is to have $x = 0$, we limit ourselves to n-th roots of non-zero numbers.

DeMoivre's Theorem. *Let n be a positive integer. For each integer k, define*

$$Z_k = \exp(2 \cdot \pi \cdot i \cdot k/n)$$

Then $Z_a = Z_b$ if and only if $a \equiv b \mod n$. The distinct Z_k are the n complex numbers whose n-th power is 1. If w is a non-zero complex number, whose polar representation is $|w| \cdot \exp(i \cdot \beta)$, then there are exactly n complex numbers whose n-th power is w, and these numbers are $|w|^{1/n} \cdot \exp(i \cdot \beta/n) \cdot Z_k$, for $k \in \mathbb{Z}_n$.

Proof. Observe that $Z_0 = 1$ and that $Z_a \cdot Z_b = Z_{a+b}$, for all integers a, b. The identity (1.1) shows that $Z_{ab} = Z_a^b$, for all integers a, b.

Each Z_k is an n-th root of 1:

$$Z_k^n = Z_{kn} = \exp(2\pi i k) = 1$$

Also, if $a \equiv b \mod n$, so that $a = b + kn$ for some integer k, then we have

$$Z_a = Z_{b+kn} = Z_b \cdot Z_{kn} = Z_b \cdot 1 = Z_b$$

Conversely, suppose that $z \in \mathbb{C}$ and $z^n = 1$. Writing z in polar form: $z = |z| \cdot \exp(i \cdot \theta)$, for some $\theta \in \mathbb{R}$, we have

$$1 = |z|^n \cdot \exp(i \cdot n\theta)$$

Looking at the modulus, $|z|^n = 1$, and since $|z| > 0$, we have $|z| = 1$. Then the definition of the exponential function shows that

$$\cos(n\theta) = 1 \quad \text{and} \quad \sin(n\theta) = 0$$

so that there is an integer k with $n\theta = 2\pi k$. We see that $z = Z_k$.

Now we show that if $Z_a = Z_b$, then $a \equiv b \bmod n$. Indeed,

$$1 = Z_a \cdot (Z_b^{-1}) = Z_a \cdot Z_{-b} = Z_{a-b} = \exp(2\pi i (b - a)/n)$$

There must be an integer k such that

$$2\pi(a - b)/n = 2\pi \cdot k \quad \text{so that} \quad a - b = kn$$

as needed. We have proved the claims about roots of 1.

Now let $w = |w| \cdot \exp(i \cdot \beta)$ be a non-zero complex number. Define $c = |w|^{1/n} \cdot \exp(i \cdot \beta/n)$, and observe that

$$c^n = |w| \cdot \exp(i \cdot \beta) = w$$

Since $c \neq 0$, there are exactly n distinct complex numbers $c \cdot Z_k$ for $k \in \mathbb{Z}_n$, and each one is an n-th root of w, since the Z_k are n-th roots of 1.

Suppose that z is an n-th root of w. Then z/c is an n-th root of 1, and so $z/c = Z_k$ for some $k \in \mathbb{Z}_n$. Thus, $z = c \cdot Z_k$. This proves that w has exactly n distinct n-th roots, represented as claimed. $\square$

You may have seen the binomial coefficients, associated with *Pascal's Triangle*; here are the properties we will need, developed from scratch. For a non-negative integer n, we define the *binomial coefficients* by this formula:

$$(1.2) \qquad \binom{n}{k} = \frac{n!}{k! \cdot (n - k)!} \quad \text{for} \quad 0 \leq k \leq n$$

The definition of the coefficients makes it obvious that they are rational numbers. You probably know that they are, in fact, integers. This can be proved by establishing the following recursive equations that define Pascal's Triangle:

$$\binom{n}{0} = 1 = \binom{n}{n}$$

and for $n \geq 1$,

$$\binom{n}{k} = \binom{n-1}{k-1} + \binom{n-1}{k} \quad \text{for} \quad 1 \leq k \leq n-1$$

One way that these coefficients come up is via the *Binomial Theorem*, which says that if a, b are numbers, then

$$(a+b)^n = \sum_{k=0}^{n} \binom{n}{k} \cdot a^k \cdot b^{n-k}$$

This can be proved by induction on n using the recursive equations.

We will need the following.

PROPOSITION 1.2. *Let p be a positive integer prime. Then $\binom{p}{k}$ is divisible by p for $1 \leq k \leq p-1$.*

PROOF. The equations (1.2) show that $\binom{p}{1} = p$. Assume that $\binom{p}{k}$ is divisible by p and that $1 \leq k \leq p-2$. Then (1.2) leads to this

$$(k+1) \cdot \binom{p}{k+1} = (k+1) \cdot \frac{p!}{(k+1)!(p-k-1)!} = \frac{p!}{k!(p-k-1)!}$$

$$= (p-k) \cdot \frac{p!}{k!(p-k)!} = (p-k) \cdot \binom{p}{k}$$

Since $1 \leq k \leq p-2$, we see that $2 \leq k+1 \leq p-1$, and so p does not divide $k+1$. Since p is prime and divides $(k+1) \cdot \binom{p}{k+1}$, we see that p divides $\binom{p}{k+1}$. $\qquad\square$

2. Finite Groups

There are certain facts of group theory that will be needed later in the course, and since the amount of group theory covered in the Foundations of Algebra course depends on the instructor, it is not clear what we can assume.

Therefore, we will begin at a fairly basic level, moving with some speed until we reach our major theorems. We include this material here, even though we may not discuss parts of it until just before those parts are needed. For the topics covered before the solvable groups, discussion and complete proofs are available in [**11**].[3]

A *finite group* is a finite set G on which there is an associative operation, usually denoted by a multiplication symbol, for which G has an identity element 1_G (often abbreviated 1), and such that each element x of G has an inverse x^{-1}, so that

$$x \cdot x^{-1} = 1_G = x^{-1} \cdot x$$

We say that the group is a group *under* the operation.

Typical group operations are not commutative: we often have $a \cdot b \neq b \cdot a$. When the group operation is commutative, the group is said to be *abelian*.

Examples

(1) The set $\mathbb{S}_n$ of permutations[4] on the set $\mathbb{Z}_n$ under function composition. The identity element is the identity function.

(2) For each positive integer prime p, the set $\mathrm{GL}(n, p)$ of $n \times n$ invertible[5] matrices with entries in $\mathbb{Z}_p$, under matrix multiplication. The identity is the $n \times n$ identity matrix I_n.

(3) The set U_n of elements of $\mathbb{Z}_n$ that have a multiplicative inverse. The identity element is 1, as an element of $\mathbb{Z}_n$.

For a finite group G, a *subgroup* is a subset which is a group under the same operation as G. If H is a non-empty subset of G and if $x, y \in H$ imply

---

[3]Boldface numbers in square brackets refer to entries in the bibliography at the end of the book, just before the index.

[4]Recall that a *permutation* is a one to one, onto function from a set to itself. The numbers $1, 2, \ldots, n$ are usually taken to represent $\mathbb{Z}_n$ in this context.

[5]A matrix in this context is invertible if its determinant is non-zero in $\mathbb{Z}_p$, or if its rank is n, performing the arithmetic of elementary operations mod p.

that $x \cdot y \in H$, then H is a subgroup. For instance, if $h \in H$, then the fact that $h^{-1} \in H$ follows from the closure of H under the group operation.

Examples

(1) If G is a finite group and $x \in G$, then the set $\langle x \rangle$ of integer powers of x is a subgroup. The order of this subgroup is the *order* of x in G: the smallest positive integer n such that $x^n = 1_G$.

(2) The set $\mathrm{SL}(n, p)$ of $n \times n$ matrices over $\mathbb{Z}_p$ that have determinant 1 is a subgroup of $\mathrm{GL}(n, p)$.

(3) the set of upper triangular elements of $\mathrm{GL}(n, p)$ is a subgroup of $\mathrm{GL}(n, p)$.

(4) Let $j \in \mathbb{Z}_n$. The set of elements f of $\mathbb{S}_n$ such that $f(j) = j$ is a subgroup of $\mathbb{S}_n$.

A group $\langle x \rangle$, generated by a single element is *cyclic*. A group is cyclic if it has the form $\langle x \rangle$ for at least one of its elements x. Cyclic groups are abelian. Not all abelian groups are cyclic. For instance, the set

$$K = \big\{ [1],\ [12][34],\ [13][24],\ [14][23] \big\} \subset \mathbb{S}_4$$

is an abelian, non-cyclic group.[6]

If H is a subgroup of the finite group G and if $x \in G$, then the set $x \cdot H$, consisting of $x \cdot h$ for all $h \in H$, is called a *left coset*. We have $|x \cdot H| = |H|$ for all $x \in G$. Distinct cosets are disjoint. These facts prove the following fundamental theorem.

LAGRANGE'S THEOREM. *If H is a subgroup of the finite group G, then $|H|$ divides $|G|$. In fact, G is the disjoint union of the distinct left cosets of H in G.*

---

[6]We will usually write permutations in the *cycle notation* used here.

The set of left cosets of H in G is written G/H. Lagrange's Theorem asserts that $|G| = |H| \cdot |G/H|$.

If G is a finite group and $x \in G$, then the order of x divides $|G|$, since the order of x is the number of elements of the subgroup $\langle x \rangle$.

Here is another basic theorem. There are several elementary proofs; we will not give one here.

CAUCHY'S THEOREM. *Let G be a group and suppose that $|G|$ is divisible by the positive integer prime p. Then G has an element of order p.*

The deeper theorems in group theory employ special subgroups that are *normal*. The word *normal* is ironic, since normal subgroups are rare, generally speaking. A subgroup is normal if it has one of the properties listed in the following proposition. The proposition states that the properties are equivalent – to prove that a given subgroup is normal, it suffices to establish any one of the properties. The notation $N \triangleleft G$ means that N is a normal subgroup of G.

PROPOSITION 1.3. *Let N be a subgroup of the finite group G. Then the following are equivalent.*

(a) if $x \in G$ and $y \in N$, then $x^{-1} \cdot y \cdot x \in N$;
(b) if $x \in G$ and $y \in N$, then $x \cdot y \cdot x^{-1} \in N$;
(c) if $x \in G$, then $xN = Nx$;
(d) if $x, y \in G$, then $xN \cdot yN = (xy)N$;
(e) the left cosets of N in G form a group under coset multiplication.

PROOF. Assume (a). Let $x \in G$ and $y \in N$, and apply (a) to x^{-1}:

$$x \cdot y \cdot x^{-1} = (x^{-1})^{-1} \cdot y \cdot (x^{-1}) \in N$$

Thus, (b) holds.

Assume (b). If $x \in G$ and $y \in N$, then (b) shows that

$$x \cdot y = x \cdot y \cdot x^{-1} \cdot x = (x \cdot y \cdot x^{-1}) \cdot x \in N \cdot x$$

This shows that $xN \subseteq Nx$. Since $|xN| = |N| = |Nx|$, we see that $xN = Nx$. This proves (c).

Assume (c). If $x, y \in G$, then

$$xNyN = x(Ny)N = x(yN)N = (xy)NN = (xy)N$$

so that (d) holds.

Assume (d). That fact shows that coset multiplication is an operation on the left cosets. It is easy to show that the associative law holds. Also, $N = 1_G \cdot N$ is the identity coset, and we have $(xN)^{-1} = (x^{-1})N$. Thus, (e) holds.

Assume (e). Let $x, y \in G$. Then $xN \cdot yN$ is a coset wN for some $w \in G$. Notice that

$$xy = x \cdot 1_G \cdot y \cdot 1_G \in xN \cdot yN$$

Thus, xy and w are in the same coset of N, and it follows that $xyN = wN$. In other words,

$$xN \cdot yN = xyN$$

Now let $x \in G$ and $n \in N$. Then

$$x^{-1} \cdot n \cdot x \in x^{-1}N \cdot xN = x \cdot x^{-1} \cdot N = N$$

This proves $x^{-1} \cdot n \cdot x \in N$, and (a) holds. $\square$

When $N \triangleleft G$, the *right* cosets of N in G also form a group, since (c) says that each right coset is really a left coset, and (d) implies that coset multiplication is essentially the same on the right as on the left. The set G/N of cosets of N in G is called a *quotient group*.

The following is straightforward.

CORRESPONDENCE THEOREM FOR GROUPS. *Let G be a finite group and $N \lhd G$.*

(a) *If H is a subgroup of G with $N \subseteq H$, then H/N is a subgroup of G/N.*

(b) *If Γ is a subgroup of G/N, then there is a subgroup H of G with $N \subseteq H$ and $H/N = \Gamma$.*

Here is a typical use.

PROPOSITION 1.4. *Let G be a finite abelian group and let m be a positive integer dividing $|G|$. Then G has a subgroup of order m.*

PROOF. Let G be a minimal counterexample. Then $m > 1$, and so m has a positive prime divisor p. We see that p divides $|G|$, and so Cauchy's Theorem gives G an element x of order p.

Since G is abelian, Proposition 1.3a holds for the subgroup $\langle x \rangle$, and so that subgroup is normal. The group $G/\langle x \rangle$ is abelian and has order less than $|G|$, and so the proposition holds for that group. If $m = p \cdot q$, since $p = |\langle x \rangle|$, Lagrange's Theorem shows that q divides $|G/\langle x \rangle|$, and so that group has a subgroup of order q. By the Correspondence Theorem, that subgroup has the form $H/\langle x \rangle$ where H is a subgroup of G. Lagrange's Theorem:

$$|H| = |\langle x \rangle| \cdot |H/\langle x \rangle| = p \cdot q = m$$

and G has a subgroup of order m. This contradiction proves the proposition. □

There is a group of order 12 that has no subgroup of order 6, and so Proposition 1.4 is not true in general.[7]

We need a few more definitions to get an important theorem. For a finite group G and $x \in G$, define $C(x)$ to be the set of $g \in G$ such that $x \cdot g = g \cdot x$.

---

[7]The alternating group $\mathbb{A}_4$ has no subgroup of order 6.

Then $C(x)$ is a subgroup of G, it is called the *centralizer of x in G*. The set $Z(G)$, the *center of the group G*, is the set of $g \in G$ with $x \cdot g = g \cdot x$ for all $x \in G$. The set $Z(G)$ is a subgroup of G; it is a subgroup of $C(x)$ for all $x \in G$, as well.

For a finite group G and $x \in G$, the *conjugacy class* of x, denoted $\mathrm{cl}(x)$, consists of $y \cdot x \cdot y^{-1}$ for all $y \in G$. There is one element of $\mathrm{cl}(x)$ for each coset of the subgroup $C(x)$ in G. In particular, $\mathrm{cl}(x) = \{x\}$ if and only if $x \in Z(G)$. Lagrange's Theorem shows that $|G| = |C(x)| \cdot |\mathrm{cl}(x)|$ for every $x \in G$.

We have $\mathrm{cl}(x) = \mathrm{cl}(y)$ if and only if $y \in \mathrm{cl}(x)$. Thus, distinct conjugacy classes are disjoint. If the distinct conjugacy classes containing more than one element are $C_1, \ldots, C_r$, then we have the *class equation*

$$|G| = |Z(G)| + \sum_{i=1}^{r} |C_i|$$

Our next theorem might be considered the first *deep* theorem of group theory. So, we will give a proof!

SYLOW'S THEOREM. *Let G be a finite group, and let q be an integer prime power dividing the order of G. Then G has a subgroup of order q.*

PROOF. Let G be a minimal counterexample. Let $q = p^m$ where p is prime and $m \geq 0$. If $m = 0$, then $q = 1$ and the subgroup $\{1_G\}$ satisfies the theorem, a contradiction. Thus, $m \geq 1$.

We consider the class equation, as written above, and we claim that p does not divide $|C_j|$ for some j. Indeed, if p divides all the $|C_j|$, then because p divides $|G|$, the class equation shows that p divides $|Z(G)|$. Cauchy's Theorem gives $Z(G)$ an element y of order p. Define $H = \langle y \rangle$, and then H is a normal subgroup of G of order p. The quotient group G/H has order less than $|G|$, and we see that p^{m-1} divides $|G/H|$. By the minimality of G, the group G/H has a subgroup A of order p^{m-1}. By the Correspondence Theorem for Groups there is a subgroup J of G such that $H \subseteq J$ and $J/H = A$. Lagrange's Theorem implies that $|J| = p \cdot |A| = p^m$, and so G has a subgroup of order p^m, a contradiction. Thus, p does not divide $|C_j|$ for some j.

Let $x \in C_j$, so that $|C_j| \cdot |C(x)| = |G|$. Since p does not divide $|C_j|$ and p^m divides $|G|$, we see that p^m divides $|C(x)|$. By definition of C_j, its order is at least 2, and so $C(x)$ is a proper subgroup of G. By the minimality of G, the group $C(x)$ has a subgroup J of order p^m. But now J is a subgroup of G of order p^m, and this is a contradiction. $\qquad\square$

3. Solvable Groups

Taking almost no time to catch our breath, we move along to the next idea. For a finite group G and $x, y \in G$, write $[x, y] = xyx^{-1}y^{-1}$. This element is called a *commutator in G*.

LEMMA 1.5. *Let H be a subgroup of the finite group G. Then $H \lhd G$ with G/H abelian if and only if $[x, y] \in H$, for all $x, y \in G$.*

PROOF. For all $x, y \in G$, compute that $xy = [x, y]yx$.

Let $H \lhd G$ and G/H abelian. If $x, y \in G$, then we have

$$Hxy = Hx \cdot Hy = Hy \cdot Hx = Hyx$$

In particular, since $xy \in xyH$, we have $xy = h \cdot yx$ for some $h \in H$. We see that $h = [x, y]$, and so every commutator in G is in H.

Converse: suppose that $[x, y] \in H$ for all $x, y \in G$. For $x \in G$ and $y \in H$, write

$$x \cdot y \cdot x^{-1} = [x, y] \cdot y \in H \cdot y = H$$

Proposition 1.3 shows that $H \lhd G$. It is easy to see that G/H is abelian: let $x, y \in G$, and since $xy = [x, y]yx$, we have

$$Hx \cdot Hy = Hxy = H[x, y]yx = Hyx = Hy \cdot Hx$$

□

For a finite group G, define G' to be the set of all finite products of commutators in G. Then G' is clearly non-empty and closed under multiplication, and so it is a subgroup of G. If $x, y \in G$, then $[x, y] \in G'$, for $[x, y]$ is a product of one commutator. Lemma 1.5 then shows that $G' \lhd G$ with G/G' abelian. Furthermore, if $H \lhd G$ and G/H is abelian, then Lemma 1.5 shows that H contains all commutators in G, and so, since H is closed under multiplication, we see that $G' \subseteq H$. The subgroup G' is called the *derived subgroup of G*.

We need the following.

LEMMA 1.6. *Let G be a finite group and $N \lhd G$. Then $N' \lhd G$.*

PROOF. Let $x, y \in N$ and let $g \in G$. Observe that

$$g[x, y]g^{-1} = [gxg^{-1}, gyg^{-1}]$$

Since $N \lhd G$, we have $gxg^{-1} \in N$ and $gyg^{-1} \in N$. Therefore, if z is a commutator in N, and if $g \in G$, then gzg^{-1} is a commutator in N.

Let $g \in G$ and $z \in N'$. Then $z = z_1 \cdots z_n$, where each z_j is a commutator in N. Then

$$gzg^{-1} = (gz_1g^{-1}) \cdots (gz_ng^{-1})$$

is a product of commutators in N. This proves that $N' \lhd G$.   □

A finite group is *solvable* if there are subgroups

(1.3) $\{1_G\} = A_0 \subseteq A_1 \subseteq A_2 \subseteq \cdots \subseteq A_k = G$

such that each A_{j-1} is a normal subgroup of A_j with A_j/A_{j-1} abelian. (By the way, it does not follow that the A_j are normal subgroups of G.) Obviously, abelian groups are solvable. So, for instance, cyclic groups are solvable.

Here are the facts we need about solvable groups.

PROPOSITION 1.7. *Let G be a solvable group. Then every subgroup of G is solvable. If N is a normal subgroup of G, then G/N is solvable. If $|G| > 1$, then G' is a proper subgroup of G.*

PROOF. Assume the notation of (1.3). If $|G| > 1$, then $|A_0| < |G|$. Let j be maximal with $|A_j| < |G|$, and then $A_{j+1} = G$ and $A_{j+1}/A_j = G/A_j$ is abelian. It follows that $G' \subseteq A_j$, and so G' is a proper subgroup of G.

If H is a subgroup of G, then you can show that

$$\{1\} = A_0 \cap H \subseteq A_1 \cap H \subseteq \cdots \subseteq A_k \cap H = G \cap H = H$$

For $1 \leq j \leq k$, let $x, y \in A_j \cap H$, and then $xyx^{-1}y^{-1} \in A_{j-1}$ since A_j/A_{j-1} is abelian, and $xyx^{-1}y^{-1} \in H$ since $x, y \in H$. Lemma 1.5 shows that

$$A_{j-1} \cap H \lhd A_j \cap H \quad \text{with} \quad (A_j \cap H)/(A_{j-1} \cap H) \quad \text{abelian}$$

Thus, H is solvable.

Assume that N is a normal subgroup of G. It will be convenient to have notation for the cosets of N. For $x \in G$, write $f(x) = xN$. Then $f : G \to G/N$ is onto, and Proposition 1.3d shows that[8]

$$f(x) \cdot f(y) = f(xy) \quad \text{for all} \quad x, y \in G$$

It is easy to see that the sets $f(A_j)$ are subgroups of $f(G) = G/N$, and we have

$$\{1_{G/N}\} = f(A_0) \subseteq f(A_1) \subseteq \cdots \subseteq f(A_k) = f(G)$$

For each j with $1 \leq j \leq k$, let $\alpha, \beta \in f(A_j)$. Then $\alpha = xN$ and $\beta = yN$ for some $x, y \in A_j$. Then $xyx^{-1}y^{-1} \in A_{j-1}$ since A_j/A_{j-1} is abelian, and so

$$\alpha\beta\alpha^{-1}\beta^{-1} = f(x) \cdot f(y) \cdot f(x^{-1}) \cdot f(y^{-1}) = f(xyx^{-1}y^{-1}) \in f(A_{j-1})$$

Lemma 1.5 implies that $f(A_{j-1}) \lhd f(A_j)$ with $f(A_j)/f(A_{j-1})$ abelian. This proves that G/N is solvable. $\square$

---

[8]You might remember that the following equation says that f is a *group homomorphism*.

The converse of Proposition1.7.

PROPOSITION 1.8. *Let G be a group, and suppose that N is a normal subgroup such that N and G/N are solvable. Then G is solvable.*

PROOF. We write the chain of subgroups that demonstrate that G/N is solvable, using the Correspondence Theorem to write each of them as a factor group. Then we see that there are subgroups of G:

$$N = B_0 \subseteq B_1 \subseteq \cdots \subseteq B_k = G$$

such that each $(B_j/N)/(B_{j-1}/N)$ is abelian. It follows that B_j/B_{j-1} is abelian. Indeed, let $x, y \in B_j$. Then $xN, yN \in B_j/N$ and so

$$xyx^{-1}y^{-1}N = xNyNx^{-1}Ny^{-1}N \in B_{j-1}/N$$

and so $xyx^{-1}y^{-1} \in B_{j-1}$. Lemma 1.5 proves that $B_{j-1} \lhd B_j$ with B_j/B_{j-1} abelian.

Since N is solvable, there are subgroups

$$\{1\} = A_0 \subseteq \cdots \subseteq A_r = N$$

such that each A_{j-1} is normal in A_j and A_j/A_{j-1} is abelian. Then $A_r = B_0$, and the chain of A_j's followed by the B_i's shows that G is solvable. $\square$

We need one more result.

PROPOSITION 1.9. *Let G be a solvable finite group with $|G| > 1$. Then there is $N \lhd G$ with N abelian and $|N| > 1$.*

PROOF. From among all $N \lhd G$ with $|N| > 1$, choose N having order as small as possible. Since $|N| > 1$, Proposition 1.7 shows that $|N'| < |N|$. By Lemma 1.6, we have $N' \lhd G$. By the minimality of N, we see that $|N'| = 1$. Because N/N' is abelian, it follows that N is abelian. $\square$

4. Problems

1. Prove (1.1) for all integers n, when $r \neq 0$. (Hint: induction!)

2. Find *nice* formulas for all the complex numbers z with $z^n = 1$ for $n = 1, 2, 3, 4, 6, 8, 16$. A *nice* formula employs integers, arithmetic operations (including division), and the square root – no trig functions. (Hint: to get from 8 to 16, use the half-angle formulas; look up those formulas, if necessary.)

3. Let G be the set of complex numbers z such that there is a positive integer n with $z^n = 1$. Show that G is a group under multiplication. (Note: it is mildly interesting that G is an infinite group in which every element has finite order.)

4. We will derive the formula for solving a cubic equation. Suppose that a_1, a_2, a_3 are complex numbers.

(a) Show that by a substitution of the form $x = y + \alpha$, the equation $x^3 + a_1 \cdot x^2 + a_2 \cdot x + a_3 = 0$ can be transformed to the form $y^3 + b_1 \cdot y + b_2 = 0$. (You will need to pick α to get the desired form.)

(b) Show that by a substitution of the form $y = z + \beta/z$, the equation $y^3 + b_1 \cdot y + b_2 = 0$ can be transformed to the form $z^6 + c_1 \cdot z^3 + c_2 = 0$. (Show how to pick β.)

Note: the quadratic formula can be applied to the transformed equation in (b) to solve for z^3; the substitutions can then be applied to find the x of the original equation. (You don't need to perform these manipulations just now.)

5. Use the method of the previous problem to find the three complex roots of $x^3 - 3x^2 - 3x + 12$. Express the roots in terms of integers, i, the arithmetic operations, and integer roots of positive real numbers. (Note: you will need to use DeMoivre's Theorem!)

6. Prove that the the binomial coefficients (defined by (1.2)) satisfy the Pascal's triangle equations.

7. Let n be a positive integer, and let $k \in \mathbb{Z}_n$ with $k \not\equiv 0$. Define $z = \exp(2\pi i k / n)$. Show that

$$\sum_{j=0}^{n-1} z^j = 0$$

(Hint: multiply by $z - 1$.)

8. Let $w = \exp(2i\pi/5)$.

    (a) Find a polynomial with integer coefficients having $w + 1/w$ as a root. (Hint: what is $(w + 1/w)^2$? Use the identity from the previous problem.)

    (b) Use the polynomial in (a) to write $w + 1/w$ in terms of integers and square roots. (You will need to pick the appropriate plus or minus sign in the quadratic formula.)

    (c) Use your formula for $w + 1/w$ to solve for w in terms of integers and square roots.

9. An integer n is a *sum of two squares* if there are integers a, b such that $n = a^2 + b^2$. Thus, $9 = 3^2 + 0^2$ and $5 = 2^2 + 1^2$ are sums of two squares. Determine, by direct calculation, which primes under 100 are sums of two squares. Make a conjecture of the form, "An integer prime is a sum of two squares if and only if..." (You are not being asked for a proof.)

10. (You will need to remember the cycle notation for permutations.) Let n be a positive integer and $x, y \in \mathbb{S}_n$. Show that $x \cdot y \cdot x^{-1}$ can be computed from the cycle structure of y by applying the permutation x to each point in the cycles of y. For instance,

$$([12][345]) \cdot [1354][26] \cdot ([12][345])^{-1} = [2435][16]$$

(Note: It follows that $\mathrm{cl}(y)$ in $\mathbb{S}_n$ is the set of elements with the same cycle structure as y.)

11. For each conjugacy class of $\mathbb{S}_4$ (for each cycle structure type), choose one element x and find $C(x)$ explicitly. Show that each coset of $C([12])$ corresponds to an element of $\mathrm{cl}([12])$.

12. Find the subgroups whose existence is predicted by Sylow's Theorem in the group $\mathbb{S}_4$.

13. Find all the elements of $\mathrm{GL}(2,2) = \mathrm{SL}(2,2)$.

14. It is a theorem that U_n is cyclic when $n = p^e$ where p is a positive odd integer prime and e is a positive integer. Find $x \in U_n$ with $\langle x \rangle = U_n$ in each case that n is an element of this set: $\{3, 5, 7, 9, 13, 25, 27\}$. (Of course, there will be a different x for each n.)

15. Let G be a finite group, $N \lhd G$, and let H be a subgroup of G. Show that HN, the set of all xy with $x \in H$ and $y \in N$, is a subgroup of G.

16. Show that $\mathbb{S}_3$ and $\mathbb{S}_4$ are solvable. (Hint: for $\mathbb{S}_4$, start with permutations of the form $[ab][cd]$.)

17. Let G be a finite group of order p^e where p is a positive integer prime and e is a positive integer. Then $|Z(G)| > 1$. (Hint: look at the class equation and show that each $|C_j|$ is divisible by p.)

18. Follow the steps to prove that if G is a finite group of order p^e where p is a positive integer prime and e is a positive integer, then G is solvable.

    (a) Let G be a minimal counterexample. Thus, G is not solvable, but every group of order equal to a prime power and less than G is solvable.

    (b) Let $z \in Z(G)$ with $z \neq 1_G$. (Why does z exist?) Then $\langle z \rangle \lhd G$. (Why?)

    (c) By the minimality of G, the group $G/\langle z \rangle$ is solvable. And $\langle z \rangle$ is abelian.

    (d) A contradiction follows.

CHAPTER 2

Rings

As with groups, coverage of ring theory in the Foundations of Algebra course varies a bit. We will introduce rings as if you have not seen them, but we will expect you to breeze through the proofs of the first few facts.

1. Ring Axioms

The paradigm example of a ring is the set of integers, with its operations: addition and multiplication. The properties of these operations are simple and easy to describe, and a moment's thought will reveal that the reader has seen other sets with operations satisfying an analogous list of properties: matrices, polynomials, as well as sets of real and complex numbers. At first we will study rings for their own sake, establishing the common properties of the examples with which we are already somewhat familiar.

Definition. A *ring* is a set R with operations $+$ and $\cdot$ satisfying the following properties, for all $a, b, c \in R$.

(1) $(a + b) + c = a + (b + c)$
(2) there is $0 \in R$
(3) for each $a \in R$ there is some $d \in R$ such that $a + d = 0$
(4) $a + b = b + a$
(5) $a \cdot (b \cdot c) = (a \cdot b) \cdot c$
(6) there is $1 \in R$ such that $0 \neq 1$ and $a \cdot 1 = a = 1 \cdot a$
(7) $a \cdot (b + c) = (a \cdot b) + (a \cdot c)$
(8) $(a + b) \cdot c = (a \cdot c) + (b \cdot c)$

Remark 1. We have used the symbols 1 and 0 in a special way; of course, since R may not be the set of integers, we do not mean the numbers 1 and 0. Momentarily we will show that the elements 1 and 0 of R are unique, and we will denote them 1_R and 0_R, whenever there is danger of confusing them with the integers of the same name.

Remark 2. Although the list mimics properties of the integers, notice that we do not assert the existence of an ordering. It is probably not obvious what is lost thereby; without going into detail at this point, we mention that we will not have the Division Theorem, in general. Another integer property that's missing: we do not necessarily have $a \cdot b = b \cdot a$ for all $a, b \in R$. Indeed, we will meet examples where this equality does not hold. On the other hand, the main rings in this course are *commutative*, meaning that $a \cdot b = b \cdot a$ for all $a, b \in R$.

Remark 3. Properties (1), (2), (3), and (4) say that the operation $+$ makes R into an abelian group. We caution that 0, not 1, is the identity element of this group. We record some immediate consequences of this group structure. As usual, we will write $-a$ for the additive inverse of a.

PROPOSITION 2.1. *Let R be a ring. Then*

(a) the element 0 mentioned in property (2) is unique;

(b) for all $a \in R$, there is a unique $-a \in R$ such that $(-a) + a = 0$;

(c) for all $a \in R$, we have $-(-a) = a$.

Remark 4. Property (5) in the definition of a ring is called the *associative law of multiplication*.

Remark 5. Property (6) identifies 1 as an identity element for multiplication. We have not said that elements have multiplicative inverses, and indeed, we will devote special study both to the case where inverses exist and to the case where they do not. Of course, not every integer has an inverse in the integers.

Remark 6. It may seem superfluous to say in property (6) that $1 \neq 0$, but we will see later that this is important. In any case, this stipulation implies that every ring has at least two elements.

Remark 7. Properties (7) and (8) are the familiar *distributive laws*.

There are several natural and elementary consequences of the definition of a ring. These are now listed, and we will use them relentlessly and without further comment.

PROPOSITION 2.2. *Let R be a ring.*

(a) *The element 1 mentioned in property (6) is unique;*

(b) *for all $a \in R$ we have $0 \cdot a = a \cdot 0 = 0$;*

(c) *for all $a \in R$ we have that*

$$(-1) \cdot a = -a = a \cdot (-1)$$

in particular, $(-1) \cdot (-1) = 1$;

(d) *for all $a, b \in R$ we have $(-a) \cdot b = -(a \cdot b) = a \cdot (-b)$;*

(e) *given $a, b \in R$ we write $a - b$ for $a + (-b)$. Then for all $a, b, c \in R$, we have*

$$-(a - b) = (b - a) \quad \text{and} \quad c \cdot (a - b) = (c \cdot a) - (c \cdot b)$$

and $(a - b) \cdot c = (a \cdot c) - (b \cdot c)$.

PROOF. If $e \in R$, and if $e \cdot a = a$ for all $a \in R$, then $e = e \cdot 1 = 1$ using property (6) for the first equality. A similar calculation shows that 1 is the only element of R satisfying $a \cdot 1 = a$ for all $a \in R$. This proves (a).

Let $a \in R$ and then $0 \cdot a = (0 + 0) \cdot a = 0 \cdot a + 0 \cdot a$ using property (8). Adding $-(0 \cdot a)$ to both sides yields $0 = 0 \cdot a$. A similar calculation using the other distributive law, property (7), proves that $a \cdot 0 = 0$.

For (c), let $a \in R$ and compute $(-1) \cdot a + a = (-1) \cdot a + 1 \cdot a = (-1+1) \cdot a$, using properties (6) and (8). Then $(-1+1) \cdot a = 0 \cdot a$ and by (b), it follows that $0 \cdot a = 0$. We conclude that $(-1) \cdot a + a = 0$. By uniqueness of additive inverses, this proves that $(-1) \cdot a = -a$. In particular, $(-1)^2 = -(-1) = 1$.

For (d), let $a, b \in R$ and compute $(-a) \cdot b = ((-1) \cdot a) \cdot b$ by (c), and $((-1) \cdot a) \cdot b = (-1) \cdot (ab)$ by property (5), and $(-1) \cdot (ab) = -(ab)$ by (c). We conclude that $(-a)b = -(ab)$. The other equalities are derived in similar fashion.

Conclusion (e) is left to the reader.   □

We will give a few examples of rings, some preliminary definitions and elementary facts, followed by more examples. For each integer $n \geq 2$, the set $\mathbb{Z}_n$ is a commutative ring, as are the integers, rational numbers, real numbers, and complex numbers. We will show how to construct many other rings inside the set of complex numbers; we will call such rings *number rings*. This situation invites the natural question when it is that a subset of a ring is also a ring, under the same operations as the larger ring. Actually, we want a little bit more: the subset S of the ring R is *subring* of R if S is a ring under the operations of R, and if $1 \in S$. The condition that $1 \in S$ can be relaxed; it fits well into the occurrence of subrings in this course.

PROPOSITION 2.3. *Let R be a ring, and let S be a subset of R. Then S is a subring of R if and only if*

(a) $1 \in S$;

(b) *if $x, y \in S$, then $x - y \in S$ and $x \cdot y \in S$.*

PROOF. If S is a subring, then (a) holds. Furthermore, if $x, y \in S$, then property (3) in the definition of ring says that y has an additive inverse in S. Since this additive inverse is unique in R, it must be $-y$. Thus, $-y \in S$. Since

S is closed under addition, we see that $x - y \in S$. Since S is closed under multiplication, we also have $x \cdot y \in S$.

For the converse, assume that (a) and (b) hold for S. We need to show that S is a ring under the operations of R.

Since $1 \in S$ and because of (b), we see that $0 = 1 - 1 \in S$. Now if $y \in S$, then $-y = 0 - y \in S$, as well. Again, if $x, y \in S$, then $-y \in S$, and so $x + y = x - (-y) \in S$. This shows that S is closed under the addition of R. Statement (b) shows that S is closed under the multiplication of R, as well.

Since addition and multiplication on R satisfy properties (1), (2), (4), (5), (6), (7), (8), the operations satisfy these properties on S. We already established property (3) in S, and we are done. $\square$

The *matrix rings* form a large class of non-commutative rings. Let R be a given ring (you can start with the integers, if you wish), and define $\mathfrak{M}(2, R)$ to be the set of 2×2 matrices with entries in R. You probably know how to define addition and multiplication on matrices:

$$\begin{pmatrix} a_{1,1} & a_{1,2} \\ a_{2,1} & a_{2,2} \end{pmatrix} + \begin{pmatrix} b_{1,1} & b_{1,2} \\ b_{2,1} & b_{2,2} \end{pmatrix}$$
$$= \begin{pmatrix} a_{1,1} + b_{1,1} & a_{1,2} + b_{1,2} \\ a_{2,1} + b_{2,1} & a_{2,2} + b_{2,2} \end{pmatrix}$$
$$\begin{pmatrix} a_{1,1} & a_{1,2} \\ a_{2,1} & a_{2,2} \end{pmatrix} \cdot \begin{pmatrix} b_{1,1} & b_{1,2} \\ b_{2,1} & b_{2,2} \end{pmatrix}$$
$$= \begin{pmatrix} a_{1,1}b_{1,1} + a_{1,2}b_{2,1} & a_{1,1}b_{1,2} + a_{1,2}b_{2,2} \\ a_{2,1}b_{1,1} + a_{2,2}b_{2,1} & a_{2,1}b_{1,2} + a_{2,2}b_{2,2} \end{pmatrix}$$

In these definitions, the $a_{i,j}$ and $b_{i,j}$ are elements of R, and the addition and multiplications that occur in the matrices indicate the operations of R. It is tedious but straightforward to prove that the operations on the matrices just given make $\mathfrak{M}(2, R)$ into a ring. Such a ring is never commutative, as is seen by multiplying the following matrices:

$$\begin{pmatrix} 0 & 1 \\ 0 & 0 \end{pmatrix} \qquad \begin{pmatrix} 0 & 0 \\ 1 & 0 \end{pmatrix}$$

We will concentrate more on commutative rings than on non-commutative rings in this course, but the matrix rings are fundamental and we will keep them in mind.

Here is an example of a subset of a ring that is a ring in its own right but not a subring of the larger ring. Consider the following set of elements of $\mathfrak{M}(2, \mathbb{Z})$:

$$\left\{ \begin{bmatrix} a & 0 \\ 0 & 0 \end{bmatrix} \;\middle|\; a \in \mathbb{Z} \right\}$$

For all $a, b \in \mathbb{Z}$, we compute that

$$\begin{bmatrix} a & 0 \\ 0 & 0 \end{bmatrix} + \begin{bmatrix} b & 0 \\ 0 & 0 \end{bmatrix} = \begin{bmatrix} a + b & 0 \\ 0 & 0 \end{bmatrix}$$

and

$$\begin{bmatrix} a & 0 \\ 0 & 0 \end{bmatrix} \cdot \begin{bmatrix} b & 0 \\ 0 & 0 \end{bmatrix} = \begin{bmatrix} a \cdot b & 0 \\ 0 & 0 \end{bmatrix}$$

This shows that our set is similar enough to the integers that it is obvious that it is a ring. The set does not contain the identity matrix for $\mathfrak{M}(2, \mathbb{Z})$, and so it is not a subring.

We now introduce other important properties that some rings have and others do not. A ring R is a *domain* if it is commutative and if $a, b \in R$ and $a \cdot b = 0$ imply that $a = 0$ or $b = 0$. For instance, $\mathbb{C}$ is a domain and every subring of $\mathbb{C}$ is too. On the other hand, in $\mathbb{Z}_4$ we have $2 \cdot 2 \equiv 0$, so that $\mathbb{Z}_4$ is not a domain. This last example is suggestive.

PROPOSITION 2.4. *The ring $\mathbb{Z}_n$ is a domain if and only if n is prime.*

PROOF. This proof should furnish a review of the properties of $\mathbb{Z}_n$. Let n be a prime, choose $a, b \in \mathbb{Z}_n$ with $a \cdot b \equiv 0$, and we must prove that $a \equiv 0$ or $b \equiv 0$.

Since $a \cdot b \equiv 0$, we have that n divides $a \cdot b$. Since n is prime, we conclude that n divides a or that n divides b. That is to say exactly that $a \equiv 0$ or $b \equiv 0$. This proves that $\mathbb{Z}_n$ is a domain.

Now suppose that n is not a prime. Then $n = a \cdot b$ where a and b are integers, each of which is greater than 1 and less than n. But then a and b are nonzero elements of $\mathbb{Z}_n$ with $a \cdot b \equiv 0$, and so $\mathbb{Z}_n$ is not a domain in this case. $\qquad\square$

To introduce our next idea, we need a notation for adding a ring element to itself over and over. For an element r of a ring R and for a non-negative integer n, we define $n * r$ to be $r + r + \cdots + r$ where there are n copies of r in the sum.[1] We leave it to you to prove that $(n+m) * r = (n * r) + (m * r)$ and that $(m * r) \cdot (n * s) = (mn) * (rs)$ for all non-negative integers n, m and all $r, s \in R$. We have to be careful with this notation, since the integer n is not necessarily an element of the ring R.

In $\mathbb{Z}_n$, it is easy to see that $n * 1 \equiv 0$, since n copies of 1 add up to n and $n \equiv 0$. In group-theoretic language, the element 1 of the abelian group $\mathbb{Z}_n$ has *finite order*. In a domain, this finite order has to be prime.

PROPOSITION 2.5. *Let R be a domain and assume that the identity element 1_R of R has finite additive order n. Then n is a prime. If $r \in R$, then $n * r = 0_R$.*

PROOF. By the definition of the order of an element in a group, the number n is the minimal positive integer such that $n * 1_R = 0_R$. Since $0_R \neq 1_R$ in a ring, we see that $n * 1_R \neq 1_R$, and so $n \neq 1$. Thus, $n > 1$, and so n has a positive prime divisor p. Write $n = p \cdot q$ for a positive integer $q < n$. Then

---

[1] This is exponentiation in the group R under $+$, but that's probably not all that helpful. To be more formal, we can use recursion to define $0 * r = 0_R$, and we define $(n + 1) * r = (n * r) + r$ for integers $n \geq 0$.

$$0_R = n * 1_R = (pq) * 1_R = (q * 1_R) \cdot (p * 1_R)$$

Since R is a domain, either $p * 1_R = 0$ or $q * 1_R = 0$. If $q * 1_R = 0$, then the minimality of n is contradicted. Thus, $p * 1_R = 0$, and the minimality of n shows that $n = p$, so that n is prime.

For $r \in R$, we have

$$n * r = (n \cdot 1) * (1_R \cdot r) = (n * 1_R) \cdot (1 * r) = 0_R \cdot r = 0_R$$

$\square$

Under the circumstances of Proposition 2.5, the prime number n is called the *characteristic* of R. Notice that the characteristic adds *every* element of R to 0. When the multiplicative identity 1 of a ring R does not have finite additive order, it is customary to say that the characteristic is *zero*. If p is a positive integer prime, then the domain $\mathbb{Z}_p$ has characteristic p. Every subring of $\mathbb{C}$ has characteristic 0.

Every non-zero rational number has a multiplicative inverse that is rational. The only integers that have an *integer* multiplicative inverse are ± 1. In general, if R is a ring (commutative or non-commutative), and if $a \in R$, then a has a *multiplicative inverse* if there is $b \in R$ such that $a \cdot b = 1 = b \cdot a$.

PROPOSITION 2.6. *Let R be a ring. Let S be the set of $a \in R$ such that a has a multiplicative inverse. Then S is a group using the ring multiplication as operation. In particular, if $a \in R$ has a multiplicative inverse, then it has a unique multiplicative inverse.*

PROOF. Observe that $1 \cdot 1 = 1$, and so $1 \in S$, thus S is not empty. If $a, b \in S$, then let c be an inverse of a, let d be an inverse of b, and compute using property (5) that

$$(a \cdot b) \cdot (d \cdot c) = a \cdot (b \cdot d) \cdot c = a \cdot 1 \cdot c = a \cdot c = 1$$

and similarly, $(d \cdot c) \cdot (a \cdot b) = 1$. This shows that $a \cdot b \in S$, and so the ring multiplication is an operation on S. Property (5) shows that the operation is associative. Property (6) shows that 1 is an identity element. Given $a \in S$, the inverse b for a is in S, since a is the inverse for b. The uniqueness of inverses follows from group theory. This completes the proof. □

If $a \in R$ has a multiplicative inverse, we write a^{-1}, as usual, for the inverse. The elements of R which have a multiplicative inverse are called *units*, and the group S defined in Proposition 2.6 is the *group of units* in R. Recall that U_n is the set of units of $\mathbb{Z}_n$. It is an elementary fact that $m \in U_n$ if and only if m is an integer with the GCD of m, n equal to 1. Thus, when n is prime, the units are all the non-zero elements of $\mathbb{Z}_n$.

The group of units in $\mathbb{Z}$ is $\{1, -1\}$. In the rational numbers, real numbers, and complex numbers all the nonzero elements have multiplicative inverses. We say that the ring R is a *field* if it is commutative and if all its nonzero elements have multiplicative inverses.

PROPOSITION 2.7. *A field is a domain. The ring R is a field if and only if the nonzero elements of R form an abelian group under multiplication.*

PROOF. Let R be a field and let $a, c \in R$ with $c \cdot a = 0$. If a is not zero then there is $b \in R$ with $a \cdot b = 1$ and so

$$c = c \cdot 1 = c \cdot (a \cdot b) = (c \cdot a) \cdot b = 0 \cdot b = 0$$

and we have proved that $c = 0$. This proves that R is a domain. Furthermore, the group S of units of R is the set of all elements of R except 0. Proposition 2.6 shows that S is a group, and since R is commutative, S is abelian.

Now suppose that R is a ring whose nonzero elements S form an abelian group. Then for $a, b \in R$ we have $a \cdot b = b \cdot a$, which holds if a or b is zero by Proposition 2.2, and holds by S being abelian if a and b are not zero. Thus R is commutative.

Given $a \in R$ with a not zero, we have that $a \in S$, hence a has an inverse b, so that $a \cdot b = 1$. This proves that R is a field.   □

Proposition 2.7 says that every field is a domain. The integers is an example of a domain that is not a field. On the other hand, a finite domain is a field:

PROPOSITION 2.8. *Let R be a domain of finite order. Then R is a field.*

PROOF. We must show that each nonzero element of R has a multiplicative inverse. Let $a \in R$ be nonzero, and define the set

$$T = \{a \cdot b \mid b \in R\}$$

and a function $f : R \to T$ by

$$f(b) = a \cdot b \quad \text{for} \quad b \in R$$

Then f is clearly onto T. We claim that f is one to one. Indeed, let $b, c \in R$ with $f(b) = f(c)$. Then $a \cdot b = a \cdot c$, so that $a \cdot b - a \cdot c = 0$, and this gives $a \cdot (b - c) = 0$. But R is a domain and a is not zero; therefore $b - c = 0$ which forces that $b = c$. Hence f is one to one.

We have that f is one to one and onto, and so R and T, being finite, have the same number of elements. But T is a subset of R, and this proves that $R = T$. In particular 1 must be an element of T, and this shows that $1 = a \cdot b$ for some $b \in R$. We have found a multiplicative inverse for a, and thus R is a field.   □

Why, exactly, doesn't Proposition 2.8 prove that the integers is a field? Don't just say, "Because the integers are not finite." Examine the proof carefully to see what doesn't work.

COROLLARY 2.9. *The ring $\mathbb{Z}_n$ is a field if and only if n is a prime.*

PROOF. Because all the $\mathbb{Z}_n$ are finite, Proposition 2.7 and Proposition 2.8 say that $\mathbb{Z}_n$ is a field if and only if it is a domain. Proposition 2.4 then settles the question. $\qquad\square$

We continue with more examples of rings, beginning with an example that will be generalized extensively. Define

$$\mathbb{Q}[\sqrt{2}] = \{a + b \cdot \sqrt{2} \mid a, b \in \mathbb{Q}\}$$

We see that $\mathbb{Q}[\sqrt{2}]$ is a subset of the real numbers. You can use Proposition 2.3 to show that this set is a subring of the real numbers. It is an exercise to show that it is a field!

Instead of using rational coefficients, we can use integer coefficients:

$$\mathbb{Z}[\sqrt{2}] = \{a + b \cdot \sqrt{2} \mid a, b \in \mathbb{Z}\}$$

This set is a subring of $\mathbb{Q}[\sqrt{2}]$. It is easy to see that it is not a field.

In working with the two rings just given, we see that the only property of $\sqrt{2}$ we use is that its square is 2. Here is a similar example over a modular set. Let α be an indeterminate,[2] and define

$$\mathbb{Z}_3[\alpha] = \{a + b \cdot \alpha \mid a, b \in \mathbb{Z}_3\}$$

We define operations on $\mathbb{Z}_3[\alpha]$, based on the desire to have $\alpha^2 = 2$. In order to distinguish the operations we are defining from those of $\mathbb{Z}_3$, we will use $\oplus$ for the addition we are defining and $\odot$ for the multiplication. We will continue to use $\equiv$ for equality, since this symbol is used in $\mathbb{Z}_3$.

$$(a + b \cdot \alpha) \oplus (c + d \cdot \alpha) \equiv (a + c) + (b + d) \cdot \alpha$$
$$(a + b \cdot \alpha) \otimes (c + d \cdot \alpha) \equiv (a \cdot c + 2 \cdot b \cdot d) + (a \cdot d + b \cdot c) \cdot \alpha$$

Compute that

$$(0 + 1 \cdot \alpha) \otimes (0 + 1 \cdot \alpha) = 2 + 0\alpha$$

and this looks like $\alpha^2 = 2$.

---

[2] An *indeterminate* is a formal symbol. The use of α here is defined by the operations.

It is tedious but straightforward to show that these operations turn $\mathbb{Z}_3[\alpha]$ into a commutative ring. Later in the course we will have an easier (more abstract!) way of doing this, so, for now, we will take it as an assumption that $\mathbb{Z}_3[\alpha]$ is a commutative ring under the given operations. Furthermore, having denoted the operations $\oplus$ and $\otimes$ to call attention to them, we now drop those symbols, reverting to the usual notation for addition and multiplication. The arithmetic on $\mathbb{Z}_3[\alpha]$ is determined by the equation $\alpha^2 = 2$. We leave it to you to show that $\mathbb{Z}_3[\alpha]$ is a field.

One more such example. Define $\mathbb{Z}_7[\alpha]$ to be the set of $a + b \cdot \alpha$ where $a, b \in \mathbb{Z}_7$, and define the operations exactly as we did for $\mathbb{Z}_3[\alpha]$, again using that $\alpha^2 = 2$. This time, $\mathbb{Z}_7[\alpha]$ is a commutative ring, but it is *not a field*, since, $(3 - \alpha) \cdot (3 + \alpha) = 0$. What's going on here?

2. Ideals

One of the fundamental ideas in group theory is that of the normal subgroup (which leads to the idea of quotient group). Kummer proffered a fundamental insight by discovering, in the mid-1800's, the analogous topic in ring theory. Kummer's papers on the subject are mentioned in [4] and [5].

A ring R is an abelian group under $+$, and so each of its subgroups is normal. A coset of the subgroup I of R looks like $x + I$ where $x \in R$. The cosets of a normal subgroup form a group under *coset addition*: given $x, y \in R$, we have $(x + I) + (y + I) = (x + y) + I$. The identity element of this group is the *identity coset* $0 + I = I$, and the additive inverse of $x + I$ is $-x + I$. We use the notation R/I, exactly as in group theory, for the set of cosets.

Since R is a ring, it is natural to try to define multiplication on R/I. Let's see if we can have

$$(2.1) \qquad\qquad (x + I) \cdot (y + I) = xy + I \quad \text{for all} \quad x, y \in R$$

This is analogous to the way addition is defined on R/I. It turns out that (2.1) does not always define an operation.[3] However, if I is what is called an *ideal*, then (2.1) defines a multiplication on R/I. Here is the definition: the additive subgroup I of R is an *ideal* if for all $x \in I$ and $a \in R$ the products xa and ax are in I.

Observe that if R is commutative, then the products xa and ax are automatically equal, and so, for commutative rings the condition is this: for all $x \in I$ and $a \in R$ we have that $xa \in I$.

We will eventually show for an ideal I of R that (2.1) defines a multiplication. First, however, in order to gain familiarity with ideals, we will find the ideals of many of the rings we already know. We start with the integers; the key idea is the Division Theorem (the quotient/remainder theorem). The proof of Theorem 2.10 is very important, because we will use analogous "quotient/remainder theorems" to prove similar results later for other rings.

THEOREM 2.10. *Let n be an integer. Then the set of integer multiples of n is an ideal of $\mathbb{Z}$. If I is an ideal of $\mathbb{Z}$, then I is the set of integer multiples of some $n \in \mathbb{Z}$.*

PROOF. For $n \in \mathbb{Z}$, we will denote by $n\mathbb{Z}$ the set of multiples of n. We leave it to you to prove that $n\mathbb{Z}$ is an ideal.

Let I be an ideal of $\mathbb{Z}$. Then I is a subgroup and so $0 \in I$. If $I = \{0\}$, then $I = 0\mathbb{Z}$ and we are done in this case.

Assume that I contains a nonzero element a. Since I is a subgroup, $-a$ is also an element of I, and thus I contains a positive element. By well-ordering there is a minimal positive element n of I, and we claim that $n\mathbb{Z} = I$.

Indeed, since $n \in I$ and I is an ideal, we have $nx \in I$ for all $x \in \mathbb{Z}$. Thus $n\mathbb{Z} \subseteq I$. Let $m \in I$ and by the Division Theorem, we can write $m = q \cdot n + r$,

---

[3]We can have $x + I = z + I$ but $xy + I \neq zy + I$, so that the "product" is not defined. Example: $R = \mathbb{Q}$, $I = \mathbb{Z}$, $x = 1$, $z = 2$, $y = 1/2$.

where $q, r \in \mathbb{Z}$ and $0 \leq r < n$. Then $q \cdot n \in I$ and $m \in I$ and so, again using that I is a subgroup, we see that

$$r = m - q \cdot n$$

is an element of I. This contradicts the minimality of n unless $r = 0$, in which case $m = q \cdot n \in n\mathbb{Z}$ This proves that $I \subseteq n\mathbb{Z}$, and the proof is complete. $\square$

If R is a commutative ring and $a \in R$, then the set

$$aR = \{x \cdot a \mid x \in R\}$$

is an ideal of R, exactly as in the case of the integers. Such an ideal is called a *principal ideal*, and one way to state Theorem 2.10 is to say that all ideals of $\mathbb{Z}$ are principal. Many, but not all, of the rings we will encounter will have the property that all their ideals are principal.

Moving back to our search for ideals, we consider the rationals, reals, and complex numbers, indeed any field.

PROPOSITION 2.11. *If R is a commutative ring, then R is a field if and only if the only ideals of R are $\{0\}$ and R itself.*

PROOF. It is easy to verify that R and $\{0\}$ are ideals of every ring R. Let R be a field, and let I be an ideal not equal to $\{0\}$. Then I has a nonzero element a, and since R is a field, a has a multiplicative inverse a^{-1}. Let $b \in R$, and then because I is an ideal, $aa^{-1}b \in I$, and hence $b \in I$. This shows that $R \subseteq I$, and thus $I = R$.

For the converse, assume that $\{0\}$ and R are the only ideals of the commutative ring R. Let $a \in R$ be nonzero, and we must show that a has a multiplicative inverse. To do this, consider the ideal aR. This ideal contains $a = 1 \cdot a$, and so it cannot be $\{0\}$. The only alternative is that $aR = R$, and therefore 1 must be an element of aR. This shows that $ab = 1$ for some $b \in R$, and the proof is complete. $\square$

It was crucial that R be commutative for the proof of Proposition 2.11. This is illustrated by our final example.

PROPOSITION 2.12. *Let F be a field. Then the only ideals of the ring $\mathfrak{M}(2, F)$ are $\{0\}$ and the ring itself.*

PROOF. Let I be an ideal of $\mathfrak{M}(2, F)$, such that I is not equal to $\{0\}$. Let $M \in I$ with M not zero, and write

$$M = \begin{pmatrix} a & b \\ c & d \end{pmatrix}$$

We claim that there is such an M with first row not zero and second row zero. Indeed, first assume that c and d are not both zero. Then

$$\begin{pmatrix} 0 & 1 \\ 0 & 0 \end{pmatrix} \cdot \begin{pmatrix} a & b \\ c & d \end{pmatrix} = \begin{pmatrix} c & d \\ 0 & 0 \end{pmatrix}$$

produces an element of I with first row not zero and second row zero. If, on the other hand, c and d are both zero, then a and b are not both zero since M is not the zero matrix. In this case M is in the desired form.

Next we claim that M can be chosen with a not zero and all other entries zero. If a is not zero, then

$$\begin{pmatrix} a & b \\ 0 & 0 \end{pmatrix} \cdot \begin{pmatrix} 1 & 0 \\ 0 & 0 \end{pmatrix} = \begin{pmatrix} a & 0 \\ 0 & 0 \end{pmatrix}$$

produces the right kind of element of I, whereas if $a = 0$, then b is not zero, and

$$\begin{pmatrix} a & b \\ 0 & 0 \end{pmatrix} \cdot \begin{pmatrix} 0 & 0 \\ 1 & 0 \end{pmatrix} = \begin{pmatrix} b & 0 \\ 0 & 0 \end{pmatrix}$$

does the job.

We now assume that M has non-zero entry only at a. Since $a \in F$ and F is a field, we have $a^{-1} \in F$.

$$\begin{pmatrix} 1 & 0 \\ 0 & 1 \end{pmatrix} = \begin{pmatrix} a^{-1} & 0 \\ 0 & 0 \end{pmatrix} \cdot \begin{pmatrix} a & 0 \\ 0 & 0 \end{pmatrix} + \begin{pmatrix} 0 & 0 \\ 1 & 0 \end{pmatrix} \cdot \begin{pmatrix} a & 0 \\ 0 & 0 \end{pmatrix} \cdot \begin{pmatrix} 0 & a^{-1} \\ 0 & 0 \end{pmatrix}$$

shows that the multiplicative identity element of the ring $\mathfrak{M}(2, F)$ is in I. It follows easily that I is the entire ring. $\qquad\square$

3. Quotient Rings

The main idea of this section is that if I is an ideal of a ring R, then R/I is itself a ring. We begin by recalling a fact from group theory: if I is an additive subgroup of R, and if $a, b \in R$, then the cosets $a + I$ and $b + I$ are equal if and only if $a - b \in I$.

If I is an ideal of R, we have already pointed out that R/I is a group under coset addition. To make R/I into a ring, we will use (2.1) to define multiplication. Since the elements of R/I are themselves sets (cosets!), we will first talk about the product of two subsets S and T of R. To avoid confusion regarding the word "product," we will use the word "join" instead. Indeed, by the *join* of S and T, we mean the set of all $s \cdot t$ where s is an element of S and t is an element of T. It is easy to show that the join operation satisfies the associative law; make sure you can state and prove that law in this context.

PROPOSITION 2.13. *Let I be an ideal of a ring R. Let a and b be elements of R, then the join of $(a + I)$ and $(b + I)$ is a subset of the coset $(a \cdot b) + I$.*

PROOF. Let x and y be elements of $a + I$ and $b + I$, respectively. Then we have $x = a + u$ for some $u \in I$, and we have $y = b + v$ for some $v \in I$. Compute that

$$x \cdot y = a \cdot b + a \cdot v + b \cdot u + u \cdot v$$

and the definition of ideal shows that the three terms av, bu, uv are in I. Thus xy is in $ab + I$. $\qquad\square$

Now we can define multiplication on R/I as in (2.1)

$$(a + I) \cdot (b + I) = ab + I$$

Proposition 2.13 shows what we are really doing: the product of two cosets is the unique coset containing their join. In the following theorem, the term *proper ideal* refers to an ideal which is not equal to the ring containing it.

THEOREM 2.14. *Let R be a ring, and let I be a proper ideal. Then R/I is a ring under coset addition and multiplication.*

PROOF. The ring properties (1)-(4) are immediate from group theory: since I is a subgroup of the abelian group R under addition, R/I is an abelian group under coset addition.

Property (5), the associative law, follows from associativity in R and the definition of multiplication in R/I.

Property (6): It is obvious that $1 + I$ is a multiplicative identity for R/I, and since I is not equal to R, $1 + I$ is not equal to the additive identity $0 + I$.

The distributive laws, properties (7) and (8), are routine. $\square$

The ring R/I is called the *quotient ring*. Notice that we have required that I not be equal to R in Theorem 2.14; this corresponds to property (6) in the definition of ring.

The most famous examples of quotient rings are the $\mathbb{Z}_n$. Indeed, if n is an integer greater than 1, then $n\mathbb{Z}$ is a proper ideal of $\mathbb{Z}$, and so the quotient ring $\mathbb{Z}/n\mathbb{Z}$ is defined. For integers a, b, the cosets $a + n\mathbb{Z}$ and $b + n\mathbb{Z}$ are equal if and only if $a - b \in n\mathbb{Z}$. In other words, the elements $a + n\mathbb{Z}$ and $b + n\mathbb{Z}$ of $\mathbb{Z}/n\mathbb{Z}$ are equal if and only if $a \equiv b \bmod n$. We see that equality in the ring $\mathbb{Z}/n\mathbb{Z}$ corresponds to equality in the ring $\mathbb{Z}_n$. On a formal level, one can *define* $\mathbb{Z}_n$ to be $\mathbb{Z}/n\mathbb{Z}$. Since, for our purposes, it seemed best to have $\mathbb{Z}_n$ around well before we were ready for quotient rings, we defined $\mathbb{Z}_n$ on its own using congruence mod n. We will continue to think of $\mathbb{Z}_n$ in the "old way," having gained some further insight by seeing that it is also a quotient ring.

Some further, intriguing examples of quotient rings are given in the problems at the end of the chapter.

The Correspondence Theorem of group theory relates the subgroups of a quotient group to those of the group. We have a direct analogy in ring theory.

CORRESPONDENCE THEOREM. *Let I be a proper ideal of the ring R.*

(a) *If $I \subseteq J \subseteq R$, where J is an ideal of R, then J/I is an ideal of R/I.*

(b) *Let $\mathcal{J}$ be an ideal of R/I. Define J to be the set of $x \in R$ such that $x + I \in \mathcal{J}$. Then J is an ideal of R with $I \subseteq J \subseteq R$ and $J/I = \mathcal{J}$.*

PROOF. The proof is exactly the same as in group theory. For (a), regarding R as an abelian group under $+$, we know from group theory that J/I is a subgroup of R/I. Furthermore, given $x \in J$, so that $x + I \in J/I$, and given $r \in R$, so that $r + I \in R/I$, we have

$$(r + I) \cdot (x + I) = r \cdot x + I$$

and since J is an ideal, we have that $rx \in J$, hence $rx + I \in J/I$. Similarly, $xr \in J$, and this long thought shows that J/I is an ideal of R/I.

For (b), group theory, again, does part of the work by showing us that J is a subgroup of R under $+$, that J contains I, and that $\mathcal{J} = J/I$. Because for $x \in I$, we have that $x + I = 0 + I = I$, we see that $I \subseteq J$. To complete the demonstration that J is an ideal of R, let $r \in R$ and $x \in J$, and then the definition of J says that $x + I \in \mathcal{J}$, so that because $\mathcal{J}$ is an ideal of R/I, we must have that

$$rx + I = (r + I)(x + I) \quad \text{and} \quad xr + I = (x + I)(r + I)$$

are elements of $\mathcal{J}$. The definition of J then shows that rx and xr are elements of J. This does it.   □

To apply the Correspondence Theorem to a given ring R, we need to be able to tell when one ideal contains another. Let's give the necessary criterion in the integers and leave it to you to supply a proof.

PROPOSITION 2.15. *Let m and n be integers. Then the ideal $n\mathbb{Z}$ of the integers contains the ideal $m\mathbb{Z}$ if and only if n divides m.*

We can use Proposition 2.15 and the Correspondence Theorem to find the ideals of each $\mathbb{Z}_n$. We already noted that we can think of $\mathbb{Z}_n$ as the quotient ring $\mathbb{Z}/n\mathbb{Z}$. The Correspondence Theorem says that the ideals of $\mathbb{Z}/n\mathbb{Z}$ come from the ideals of $\mathbb{Z}$ containing $n\mathbb{Z}$. Theorem 2.10 tells us that the ideals of $\mathbb{Z}$ are of the form $m \cdot \mathbb{Z}$, and Proposition 2.15 tells us that $n \cdot \mathbb{Z} \subseteq m \cdot \mathbb{Z}$ if and only if m divides n. We observe that $m\mathbb{Z} = -m\mathbb{Z}$, for each integer m, and so we can use positive divisors m of n to get the distinct ideals we are looking for. For each such m, we have $n\mathbb{Z} \subseteq m\mathbb{Z}$ and we get an ideal $m\mathbb{Z}/n\mathbb{Z}$ of $\mathbb{Z}/n\mathbb{Z} = \mathbb{Z}_n$. Can you see that $m\mathbb{Z}/n\mathbb{Z}$ is like $\mathbb{Z}_q$ for some q?

Later in the course, we will construct quotient rings that are fields. For now, we show how to use the ideal involved to tell whether the quotient ring is a field. Suppose that R is a commutative ring and that I is a proper ideal. Then R/I is a commutative ring, and Proposition 2.11 says that it will be a field precisely when its only ideals are the zero ideal $\{0 + I\}$ and the ring R/I itself. The Correspondence Theorem allows us to say this in terms of the ring R: ideals of R/I come from ideals of R between I and R, in a one to one correspondence. Thus, R/I will have only two ideals if and only if there are only two ideals between I and R, namely I and R. We make a definition to identify this situation: we call I a *maximal ideal* of R if I is a proper ideal of R and $I \subseteq J \subseteq R$ for an ideal J implies that $J = I$ or $J = R$.

The following is then obvious.

PROPOSITION 2.16. *Let I be a proper ideal of the commutative ring R. Then I is a maximal ideal of R if and only if R/I is a field.*

Proposition 2.16 will be used later to construct important examples of fields. For now we identify the maximal ideals of the integers. You are asked to supply a proof using Proposition 2.15.

PROPOSITION 2.17. *For an integer n, the ideal $n\mathbb{Z}$ of the integers is maximal if and only if n is a prime.*

When R is a non-commutative ring, the definition of maximal ideal still makes sense, but when I is a maximal ideal of R, the structure of R/I can be complicated. We will not pursue this except to use a ring we have already studied to give an example. Since $\mathbb{Q}$ is a field, Proposition 2.12 shows that the only proper ideal of $\mathfrak{M}(2, \mathbb{Q})$ is $\{0\}$. Thus, $\{0\}$ is a maximal ideal of $\mathfrak{M}(2, \mathbb{Q})$. The quotient ring $\mathfrak{M}(2, \mathbb{Q})/\{0\}$ has the same structure as $\mathfrak{M}(2, \mathbb{Q})$, and so it is not a field – it is not even commutative.

4. Ring Homomorphisms

Recall that a group homomorphism is a function $f : G \to H$ where G and H are groups and such that

$$f(x \cdot y) = f(x) \cdot f(y) \quad \text{for all} \quad x, y \in G$$

Note that the multiplication dot on the left side $x \cdot y$ denotes the operation of G, whereas the dot on the right $f(x) \cdot f(y)$ is the multiplication of H.

Ring homomorphisms are defined analogously, using both the operations of the ring. Indeed, if R and S are rings, then $f : R \to S$ is a *ring homomorphism* provided, for all $x, y \in R$, we have

$$f(x + y) = f(x) + f(y)$$
$$\text{and} \quad f(x \cdot y) = f(x) \cdot f(y)$$

Here we have used $+$ and $\cdot$ to stand for the operations of both rings.

Two trivial examples of ring homomorphisms are as follows: first, let $f(x) = 0$ for all $x \in R$; second, let $R = S$ and let f be the identity function. The first of these examples shows that $f(1_R)$ does not have to be 1_S.

An important, non-trivial example. Let I be a proper ideal of the ring R, and define $f : R \to R/I$ by $f(x) = x + I$ for all $x \in R$. It is an exercise to show that f is a ring homomorphism; it is called a *canonical homomorphism*.

We have already shown that $\mathbb{Z}_n$ can be identified with $\mathbb{Z}/n\mathbb{Z}$. The canonical homomorphism $f : \mathbb{Z} \to \mathbb{Z}/n\mathbb{Z}$ can be viewed as mapping onto $\mathbb{Z}_n$ if we wish.

As in our discussion of quotient rings, our knowledge of groups provides some free information, given in (a) and (b) below.

PROPOSITION 2.18. *Let $f : R \to S$ be a ring homomorphism. Then*

(a) *viewing R and S as abelian groups under $+$, f is a group homomorphism.*

(b) *We have $f(0_R) = 0_S$ and, for all $r \in R$, $f(-r) = -f(r)$.*

(c) *If $f(r) = 1_S$ for some $r \in R$ (for example, if f is onto), then $f(1_R) = 1_S$.*

(d) *If S is a domain and if $f(x) \neq 0$ for some $x \in R$, then $f(1_R) = 1_S$.*

(e) *If $1_S \in f(R)$, then $f(R)$ is a subring of S.*

PROOF. Statements (a) and (b) are immediate from the observation that f is a group homomorphism with respect to $+$.

For (c), we caution you that the r such that $f(r) = 1_S$ does not have to be 1_R. But, given such an r, we can calculate

$$1_S = f(r) = f(r \cdot 1_R) = f(r) \cdot f(1_R) = 1_S \cdot f(1_R) = f(1_R)$$

which proves (c).

If S is a domain and if $f(x)$ is not zero, then

$$(f(1_R) - 1_S) \cdot f(x) = f(1_R) \cdot f(x) - f(x) = f(x) - f(x) = 0_S$$

shows that $f(1_R) - 1_S = 0$ which proves (d). You do (e).   □

Every normal subgroup of a group is the *kernel* of a group homomorphism. Analogy: every ideal of a ring is the *kernel* of a ring homomorphism. Suppose we have a ring homomorphism $f : R \to S$. The *kernel* of f is denoted $\ker(f)$, and it is define to be the set of $x \in R$ such that $f(x) = 0_S$. We have remarked that f is a group homomorphism (of abelian groups under addition), and the kernel just defined is the kernel of the group homomorphism. In the group setting, the kernel is a subgroup; for rings, the kernel is an ideal.

PROPOSITION 2.19. *Let $f : R \to S$ be a ring homomorphism. Then $\ker(f)$ is an ideal of R. The function f is one to one if and only if $\ker(f) = \{0_R\}$.*

PROOF. For the first statement, we only need to let $x \in \ker(f)$ and $r \in R$ and show that $x \cdot r$ and $r \cdot x$ are elements of $\ker(f)$. Indeed,

$$f(xr) = f(x) \cdot f(r) = 0_S \cdot f(r) = 0_S$$

and

$$f(rx) = f(r) \cdot f(x) = f(r) \cdot 0_S = 0_S$$

The statement about when f is one to one is proved exactly as in group theory; nonetheless, we rehearse it here. If f is one to one, then only 0_R can map to 0_S, and so $\ker(f) = \{0_R\}$. On the other hand, if $\ker(f)$ only has 0_R, then let $f(x) = f(y)$ for $x, y \in R$, and observe that $f(x-y) = f(x) - f(y) = 0_S$, so that $x - y$ is in $\ker(f)$, and thus $x = y$. This shows that f is one to one.   □

Suppose that I is a proper ideal of the ring R, and let $f : R \to R/I$ be the canonical homomorphism. The kernel of f is the set of $x \in R$ such that $x + I = I$, and this set is I! Thus, every proper ideal is the kernel of a homomorphism.[4]

------

[4]The ideal $I = R$ is also a kernel. Define $f : R \to R$ by $f(x) = 0$ for all $x \in R$.

By analogy with group isomorphisms, a ring homomorphism that is one to one and onto is called a *ring isomorphism*. If there is a ring isomorphism from ring R onto ring S, then we say that R is *isomorphic* to S. As we expect, isomorphic rings are, in some sense, the "same ring." This is fleshed out in exercises at the end of the chapter. Here are two important examples.

Define C to be the set of elements of $\mathfrak{M}(2, \mathbb{R})$ of the form

$$\begin{pmatrix} a & -b \\ b & a \end{pmatrix}$$

We leave it to you to show that C is a subring. We will show that C is isomorphic to the complex numbers. To get started, define

$$J = \begin{pmatrix} 0 & -1 \\ 1 & 0 \end{pmatrix} \quad \text{and compute} \quad J^2 = \begin{pmatrix} -1 & 0 \\ 0 & -1 \end{pmatrix}$$

Writing I for the 2×2 identity matrix (the multiplicative identity element of C), we have $J^2 = -I$, and this looks like the equation $i^2 = -1$ in $\mathbb{C}$. Furthermore, notice that

$$\begin{pmatrix} a & -b \\ b & a \end{pmatrix} = a \cdot I + b \cdot J$$

and this is like $a + b \cdot i$. We define $f : C \to \mathbb{C}$ by $f(a \cdot I + b \cdot J) = a + i \cdot b$. It is a direct calculation to show that this is a ring isomorphism. Turning the tables, we could regard C as the *definition* of the complex numbers, with J as the formal definition of i. In other words, C provides a *construction* of the complex numbers.

Another example. The real numbers $\pm\sqrt{2}$ are both roots of $X^2 - 2$. Define $f : \mathbb{Q}[\sqrt{2}] \to \mathbb{Q}[\sqrt{2}]$ by $f(a + b \cdot \sqrt{2}) = a - b \cdot \sqrt{2}$. You should show that f is a ring isomorphism. Since the domain and image of f are the same, it could be called an *automorphism* as well. Notice that $f(\sqrt{2}) = -\sqrt{2}$ and $f(-\sqrt{2}) = \sqrt{2}$, so that f permutes the roots of $X^2 - 2$. We will see that permutations of the roots of a polynomial often lead to automorphisms – this is the Galois theory that will figure prominently later in the course.

To construct an isomorphism of fields, all you need is an onto homomorphism.

PROPOSITION 2.20. *Let R and S be fields and let $f : R \to S$ be an onto ring homomorphism. Then f is an isomorphism.*

PROOF. Since f is onto, there is $x \in R$ such that $f(x) = 1_S$. This proves that $x \notin \ker(f)$, and so $\ker(f) \neq R$.

Since Proposition 2.19 shows that $\ker(f)$ is an ideal, Proposition 2.11 tells us that $\ker(f)$ is either R or $\{0_R\}$. The former case has been ruled out, and so $\ker(f) = \{0_R\}$. Proposition 2.19 says that f is one to one. Thus, f is an isomorphism.   □

Now we can get an interesting (and perhaps surprising) example of a ring homomorphism. Recall the definition of the *characteristic* of a domain.

PROPOSITION 2.21. *Let R be a domain of prime characteristic p. Then the function $f : R \to R$ defined by $f(x) = x^p$ is a one to one ring homomorphism.*

PROOF. It is easy to see that $(x \cdot y)^p = x^p \cdot y^p$ for all $x, y \in R$. We use the Binomial Theorem[5] to evaluate

$$(x + y)^p = x^p + \sum_{k=1}^{p-1} \binom{p}{k} * (x^k y^{n-k}) + y^p$$

Proposition 1.2 says that p divides each $\binom{p}{k}$ in the middle sum. Since p is the characteristic of R, all these terms of 0, and we conclude that $(x+y)^p = x^p + y^p$. This shows that f is a ring homomorphism.

To see that f is one to one, let $x \in \ker(f)$, so that $x^p = 0$. Since R is a domain, we have $x = 0$. Proposition 2.19 shows that f is one to one.   □

---

[5]We are generalizing the Binomial Theorem to the setting of an arbitrary ring. It can be proved by an induction argument virtually identical to that given for the number version of the theorem. Recall the $n * r$ notation for n copies of the ring element r added.

5. Problems

1. In the definition of *ring*, omit property (4). Show that property (4) follows as a theorem from the other axioms. (Hint: expand $(1+1)(a+b)$ for $a, b \in R$.)

2. Let R be a ring in which $a^2 = a$ for all $a \in R$. Show that R is commutative.

3. Let R be a ring and suppose that $a^3 = a$ for all $a \in R$. Show that R is commutative.

4. Show that $\mathbb{Q}[\sqrt{2}]$ is a field.

5. Recall the complex number i. Define $\mathbb{Z}[i]$ similar to the way that $\mathbb{Z}[\sqrt{2}]$. Show that this set is a subring of $\mathbb{C}$ and that it is a domain.

6. For each positive integer n, define $\mathbb{Z}_n[\alpha]$ where $\alpha^2 = 2$, as was done in the case $n = 3$ and $n = 7$. For which of the following n is $\mathbb{Z}_n[\alpha]$ a field? $n = 2, 3, 5, 7, 17$

7. For each positive integer n, define $\mathbb{Z}_n[i]$ by analogy with $\mathbb{Z}[i]$, with the natural operations. For which of the following n is $\mathbb{Z}_n[i]$ a field? $n = 2, 3, 5, 7, 13$

8. Let $\alpha \in \mathbb{C}$ be a root of the polynomial $X^2 + b \cdot X + c$ where $b, c \in \mathbb{Z}$. Define $\mathbb{Q}[\alpha]$ to be the set of $r + s \cdot \alpha$ where $r, s \in \mathbb{Q}$. Show that $\mathbb{Q}[\alpha]$ is a field. (Note: one case is where $\alpha \in \mathbb{Q}$; in that case, show that $\mathbb{Q}[\alpha] = \mathbb{Q}$.)

9. Define $\mathbb{Z}_2[\alpha]$ to be the set of $a + b \cdot \alpha$ where $a, b \in \mathbb{Z}_2$, and such that $\alpha^2 = \alpha + 1$. Show that $\mathbb{Z}_2[\alpha]$ is a field.

10. Define $\mathbb{Z}_2[\alpha]$ to be the set of $a + b \cdot \alpha + c \cdot \alpha^2$, where $a, b, c \in \mathbb{Z}_2$ and such that $\alpha^3 = \alpha + 1$. Show that $\mathbb{Z}_2[\alpha]$ is a field of order 8.

11. Let R, S be rings, and define $R \oplus S$ to be the set of (r, s) with $r \in R$ and $s \in S$. Define $(a, b) + (c, d) = (a + c, b + d)$ and $(a, b) \cdot (c, d) = (ac, bd)$ for all $a, c \in R$ and $b, d \in S$. Show that this makes $R \oplus S$ into a ring.[6] The

---

[6] As a set, $R \oplus S$ is the cartesian product $R \times S$. The $\oplus$ notation is customary for rings.

ring $R \oplus S$ is called the *direct sum* of R and S. Show that $R \oplus S$ is *never* a domain.

12. Let R be a ring. Recall that $\mathbb{N}$ is the set of non-negative integers. Define $n * r$ for all $n \in \mathbb{N}$ and $r \in R$, recursively by the equations

$$0 * r = 0_R \quad \text{and} \quad (n+1) * r = (n * r) + r \quad \text{for all} \quad r \in R,\ n \in \mathbb{N}$$

Prove the following, for all $n, m \in \mathbb{N}$ and $r, s \in R$.

(a) $(n + m) * r = (n * r) + (m * r)$

(b) $n * (r + s) = (n * r) + (n * s)$

(c) $(n * r) \cdot (m * s) = (nm) * (rs)$

(Hint: induction on n or m.)

13. Show that ring isomorphism is an *equivalence relation*:

(a) The ring R is isomorphic to itself.

(b) If R is isomorphic to S, then S is isomorphic to R.

(c) If R is isomorphic to S and S to T, then R is isomorphic to T.

14. For each positive integer n, define the subset R_n of $\mathfrak{M}(2, \mathbb{Z}_n)$ to consist of matrices of the form

$$\begin{pmatrix} a & -b \\ b & a \end{pmatrix} \quad \text{where} \quad a, b \in \mathbb{Z}_n$$

Show that R_n is a subring of $\mathfrak{M}(2, \mathbb{Z}_n)$. Compute

$$\begin{pmatrix} 0 & -1 \\ 1 & 0 \end{pmatrix}^2$$

and show that R_n is isomorphic to $\mathbb{Z}_n[i]$.

15. Let n be a positive integer. Observe that

$$\begin{pmatrix} 0 & 1 \\ 2 & 0 \end{pmatrix}^2 = \begin{pmatrix} 2 & 0 \\ 0 & 2 \end{pmatrix}$$

Use this equation to find a subring of $\mathfrak{M}(2, \mathbb{Z}_n)$ isomorphic to $\mathbb{Z}_n[\alpha]$, where $\alpha^2 = 2$.

16. Show that $1 + \sqrt{2}$ is a unit in $\mathbb{Z}[\sqrt{2}]$. Find $(1 + \sqrt{2})^n$ for $-3 \le n \le 3$. (Write them in $a + b \cdot \sqrt{2}$ form.) Notice that $\mathbb{Z}[\sqrt{2}]$ has infinitely many units.

17. (This requires a little matrix theory.) Show that the units of $\mathfrak{M}(2, \mathbb{Z})$ are the matrices with determinant ± 1. Show that there are infinitely many such units.

18. Let I be an ideal of the ring R. Show that $I = R$ if and only if I has an element that is a unit.

19. Let I, J be ideals of a ring R. Show that $I \cap J$ is an ideal of R.

20. Let I, J be ideals of a ring R, and suppose that $I \cap J = \{0\}$. Show that $x \cdot y = 0$ for all $x \in I$ and $y \in J$.

21. Let I, J be ideals of a ring R. Define $I + J$ to be the set of $x + y$ for all $x \in I$ and $y \in J$. Show that $I + J$ is an ideal of R.

22. Show that $2 \cdot \mathbb{Z} + 3 \cdot \mathbb{Z} = \mathbb{Z}$.

23. If I is a subgroup under addition of a ring R, and if $x \in R$ and $y \in I$ implies that $xy \in I$, then we say that I is a *left ideal* of R. If $a \in R$, show that Ra is a left ideal of R.

24. Show that the set of

$$\begin{pmatrix} 0 & a \\ 0 & b \end{pmatrix} \quad \text{for all} \quad a, b \in \mathbb{Q}$$

is a left ideal of $\mathfrak{M}(2, \mathbb{Q})$.

25. Let $R = \mathbb{Z}[\sqrt{2}]$, and define $I = 3 \cdot \mathbb{Z}[\sqrt{2}]$. Show that R/I is a field of order 9, isomorphic to $\mathbb{Z}_3[\alpha]$, where $\alpha^2 = 2$.

26. Let R, S be rings, and let K be an ideal of the direct product $R \oplus S$.

(a) Let $(a, b) \in K$, and show that $(a, 0) \in K$ and that $(0, b) \in K$.
(b) Define I to be the set of $a \in R$ such that $(a, 0) \in K$. Show that I is an ideal of R. Similarly, define J to be the set of $b \in S$ such that $(0, b) \in K$, and J is an ideal of S.
(c) Show that $K = I \oplus J$.

27. Let A be a non-empty set and let R be a commutative ring. Define $\mathfrak{F}(A, R)$ to be the set of functions $f : A \to R$. For $f, g \in \mathfrak{F}(A, R)$, define $f + g$ by the formula $(f + g)(x) = f(x) + g(x)$ for all $x \in A$, and define $f \cdot g$ by $(f \cdot g)(x) = f(x) \cdot g(x)$ for all $x \in A$. Show that $\mathfrak{F}(A, R)$ is a commutative ring. Could it be a field?

28. Let A be a non-empty set, let R be a commutative ring, and let B be a subset of A. Let I be the set of $f \in \mathfrak{F}(A, R)$ such that $f(b) = 0_R$ for all $b \in B$. Show that I is an ideal of $\mathfrak{F}(A, R)$.

29. Let A be a non-empty finite set, and let R be a field. Let I be an ideal of $\mathfrak{F}(A, R)$. Let B be the set of $b \in A$ such that $f(b) = 0$ for all $f \in I$. Let J be the set of $f \in \mathfrak{F}(A, R)$ such that $f(b) = 0$ for all $b \in B$. Show that $I = J$, via the following steps.

(a) Show that $I \subseteq J$.
(b) For each $a \in A \setminus B$, find $f_a \in I$ such that $f_a(a) \neq 0$.
(c) Use that R is a field to adjust f_a so that $f_a(a) = 1$ and $f_a(b) = 0$ for all $b \neq a$.
(d) Show that each $g \in J$ is a sum of elements $g \cdot f_a$.

30.   Find the proper ideals of $\mathbb{Z}_{12}$ and show that each quotient ring looks the same as some $\mathbb{Z}_n$.

31.   Let R be a commutative ring in which all ideals are principal ideals. Let I be a proper ideal of R. Prove that all ideals of R/I are principal.

32.   Let I be an ideal of $\mathbb{Z}[\sqrt{2}]$ and suppose that $I \neq \{0\}$. Show that $\mathbb{Z}[\sqrt{2}]/I$ is finite, completing the following steps:

(a) Find a positive integer n in I. (Hint: if $a + b \cdot \sqrt{2} \in I$, multiply by $a - b\sqrt{2}$.)
(b) Show that every element of $\mathbb{Z}[\sqrt{2}]/I$ can be written $(a + b \cdot \sqrt{2}) + I$ where $0 \leq a < n$ and $0 \leq b < n$.
(c) Show that the quotient ring has at most n^2 elements.

33.   Let R be a commutative ring. The ideal I of R is a *prime ideal* of R if I is proper and if $x, y \in R$ and $x \cdot y \in I$ imply that $x \in I$ or $y \in I$. Show that the prime ideals of $\mathbb{Z}$ are the ideals $n\mathbb{Z}$ where n is prime.

34.   Let R be a commutative ring, and let I be an ideal of R. Show that I is a prime ideal if and only if R/I is a domain.

35.   Show that $\sqrt{2} \cdot \mathbb{Z}[\sqrt{2}]$ is a prime ideal of $\mathbb{Z}[\sqrt{2}]$.

36.   Let $f : R \to S$ be a ring isomorphism.

(a) If R is commutative, then S is commutative.
(b) If R is a domain, show that S is a domain.
(c) The function f defines an isomorphism from the units group of R onto the units group of S.
(d) If R is a field, then S is a field.
(e) The subset I of R is an ideal if and only if $f(I)$ is an ideal of S.
(f) If I is a proper ideal of R, then the ring R/I is isomorphic to the ring $S/f(I)$.

37.   Let n be a positive integer. Define R to be the set of all elements of $\mathfrak{M}(2, \mathbb{Z}_n)$ of the form

$$\begin{pmatrix} a & b \\ 3b & a \end{pmatrix}$$

Find a ring isomorphism from R onto the ring $\mathbb{Z}_n[\alpha]$ where $\alpha^2 = 3$.

38.   Find a ring R like the one in the previous problem that is ring isomorphic to $\mathbb{Q}[w]$ where $w = \exp(2\pi i/3)$.

39.   Find the smallest positive integer n such that there are two rings of order n that are *not* isomorphic. (Find example rings of order n, and also prove that n is minimal.)

40.   Let $p \in \mathbb{Z}$ where p is not the square of an integer. Let $\alpha \in \mathbb{C}$ be chosen so that $\alpha^2 = p$. We have defined the ring $\mathbb{Q}[\alpha]$ previously: the set of $a + b \cdot \alpha$ where $a, b \in \mathbb{Q}$. Define $f : \mathbb{Q}[\alpha] \to \mathbb{Q}[\alpha]$ by $f(a + b \cdot \alpha) = a - b \cdot \alpha$. Show that f is a ring automorphism. Observe that f gives an automorphism of $\mathbb{Z}[\alpha]$, as well.

41.   Recall the ring $\mathbb{Z}_3[\alpha]$ where $\alpha^2 = 2$. Show that the mapping $f(a+b\cdot\alpha) = a - b \cdot \alpha$ is a ring isomorphism.

42.   Let $f : \mathbb{Z} \to \mathbb{Z}$ be a ring automorphism. Show that f is the identify function. Do the same thing for $\mathbb{Q}$ in place of $\mathbb{Z}$.

43.   This exercise is quite abstract, but it exhibits a common level of thinking about rings. Let R be an abelian group under $+$, and define $\text{End}(R)$ to be the set of homomorphisms $f : R \to R$. Show that $\text{End}(R)$ is a ring where the addition is defined by $(f + g)(r) = f(r) + g(r)$, and the multiplication is function composition. The ring $\text{End}(R)$ is the *endomorphism ring* of R.

44.   Let R be a ring. Note that $\text{End}(R) \subset \mathfrak{F}(R, R)$, but show that $\text{End}(R)$ is not a subring of $\mathfrak{F}(R, R)$. (You may take $R = \mathbb{Z}_2$ to give a simple example.)

45. Let R be a ring. Show that the set of automorphisms of R forms a group under function composition. Show that this group is the units group of $\operatorname{End}(R)$.

46. (Recall the definition of the direct sum of rings.) This problem is meant to show you that you have already seen endomorphism rings, cleverly disguised. Let $f \in \operatorname{End}(\mathbb{Q} \oplus \mathbb{Q})$. Suppose that $f(1,0) = (a,b)$ and $f(0,1) = (c,d)$. Write

$$\alpha(f) = \begin{pmatrix} a & c \\ b & d \end{pmatrix}$$

Show that α is a ring isomorphism between $\operatorname{End}(\mathbb{Q} \oplus \mathbb{Q})$ and $\mathfrak{M}(2, \mathbb{Q})$.

47. (This fact is sometimes called the *First Isomorphism Theorem.*) Let $f : R \to S$ to an onto ring homomorphism. Then there is an isomorphism of $R/\ker(f)$ with S that sends the coset $x + \ker(f)$ in $R/\ker(f)$ to the element $f(x)$ of S.

48. (This fact is variously called the *Second Isomorphism Theorem* or the *Third Isomorphism Theorem.*) Let I be a proper ideal of the ring R, and let J be a proper ideal of R with $I \subseteq J$. The Correspondence Theorem says that J/I is a proper ideal of R/I. Show that the quotient ring $(R/I)/(J/I)$ is isomorphic to R/J.

CHAPTER 3

Factorization

Now we are ready to generalize the concept of integer factorization to other rings. In order to keep ideals and quotient rings in view, we will formulate each new concept in terms of them.

1. Units, Irreducibles, Primes

Throughout, the rings we consider will be domains – in particular, they are commutative. For such a ring R, if $m, n \in R$, we say that m *divides* n, if $n = m \cdot r$ for some $r \in R$. In terms of ideals, m divides n if and only if $mR \supseteq nR$; you can remember that "divides means contains."

We caution you that we are **not** introducing "division" as a ring operation. In performing calculations, we use multiplication and addition as before. Secondly, notice that you have to know what ring you are talking about before you can decide whether one element divides another. For example, as elements of the integers, 2 does not divide 3, but as rational numbers, $3 = 2 \cdot (3/2)$ and so 2 does divide 3. Clearly, all elements of a ring R divide 0_R. What about the elements which divide 1?

PROPOSITION 3.1. *Let R be a domain.*

(a) The elements which divide 1 are precisely the units of R.

(b) If x and y are units of R, then x divides y (and y divides x).

(c) If $x, y \in R$, if x divides y, and y is a unit, then x is a unit.

PROOF. If x divides 1, then $1 = xy$ for some $y \in R$. Since R is commutative, we also have $1 = yx$ and thus x has a multiplicative inverse, and so x is a unit.

If x and y are units, then $y = x(x^{-1}y)$, which shows that x divides y.

If $y = xr$ for some $x, y, r \in R$ and if y is a unit, then $1 = y \cdot y^{-1} = xry^{-1}$, and so x divides 1, which makes x a unit. $\square$

Proposition 3.1 says that the factors of the units are all the units. If $m, n \in R$ we say that m and n are *associates* if $m = u \cdot n$ for some unit u in R. Observe in this case that $n = u^{-1}m$ and so n and m are associates as well. How do you detect associates with ideals?

PROPOSITION 3.2. *Let R be a commutative domain, and let $m, n \in R$. Then m and n are associates if and only if $mR = nR$. In particular, the element u of R is a unit if and only if $R = uR$.*

PROOF. If m and n are associates, say $m = u \cdot n$ where u is a unit, then n divides m and so nR contains mR. But then also $n = u^{-1}m$ and so m divides n, hence mR contains nR. Thus $mR = nR$.

If $mR = nR$, then $m = nr$ and $n = ms$ where $r, s \in R$. This leads to $m = mrs$ which forces $m(1 - rs) = 0$. Since R is a domain, we either have $m = 0$ or $1 = rs$. If $m = 0$, then $n = 0$ (why?) and so $m = n$ are certainly associates. If $1 = rs$, then by Proposition 3.1a r is a unit, and again m and n are associates.

The other conclusion then follows from Proposition 3.1. $\square$

Back to factorization, what about the elements which are not zero and not units? By analogy with the integers, we expect that "primes," will be the building blocks of factorization. Recall that the integer primes have two properties:

(a) If the integer m divides the prime p then m is one of the numbers: $1, -1, p, -p$

(b) If the prime p divides the product $m \cdot n$ of integers, then either p divides m or p divides n.

Property (a) can be read, "if $p = m \cdot n$, then m is a unit or n is a unit." This statement is often used as the definition of integer prime. Statement (b) is then proved as a consequence. However, the proof of (b) involves the Division Theorem, and the rings we encounter in general will not always have a "division theorem." We will find that (b) does not follow from (a) in all cases, and we are thus led to distinguish the two properties.

If $p \in R$ is not zero and not a unit, then we say p is *irreducible* if whenever $p = m \cdot n$ for $m, n \in R$, we have that m is a unit or n is a unit. (This property is a generalization of (a).) If R is a commutative domain and if $p \in R$ is not 0 and not a unit, then we say that p is *prime* if whenever p divides $m \cdot n$ for $m, n \in R$ it must be that p divides m or p divides n. (We are generalizing (b).)

In a problem below we will give an example of an irreducible which is not a prime. However, primes are always irreducible.

PROPOSITION 3.3. *Let R be a commutative domain, and let p be a prime element of R. Then p is an irreducible.*

PROOF. Let $p = m \cdot n$, and then p divides $m \cdot n$. By definition of prime, p divides either m or n. Say, for example, that p divides m. Write $m = p \cdot r$ and then $p = p \cdot r \cdot n$, so that $p \cdot (1 - rn) = 0$. Since p is not zero (part of the definition of being prime!) we must have that n is a unit. Similarly, if p divides n, then m is a unit. $\square$

You should convince yourself that p is a prime if and only if all its associates are primes too. Similarly, p is an irreducible if and only if its associates are too.

The next theorem is the most used fact about primes, although this will probably not be obvious for a while. In the rest of this chapter we will often write $p_1 \ldots p_m$ for the product of p_j with $1 \leq j \leq m$.

THEOREM 3.4. *Let R be a commutative domain with $n \in R$, and suppose that $n = p_1 \ldots p_m$ where each p_i is an irreducible element of R. Let q be a prime element of R with q dividing n. Then q is an associate of one of the p_i.*

PROOF. We prove this by induction on m. If $m = 1$, then n is irreducible, and since q divides n and q is not a unit, it must be that q is an associate of $n = p_1$.

Now assume that the theorem is true when there are k irreducible factors, and assume n can be written in terms of $k + 1$ factors. Let $r = p_1 \ldots p_k$, and then we see that q divides $r \cdot p_{k+1}$. If q divides r, then by induction q must be an associate of some p_i. Since q is prime, the only other possibility is that q divides p_{k+1}, and then, as in the previous paragraph, q is an associate of p_{k+1}. $\square$

Notice that the associate p_i of q must be prime, not just irreducible.

One often says that each integer (except $0, -1, 1$) factors uniquely into primes. What does the word "uniquely" mean? First of all, it does not mean "in a unique order," since multiplication is commutative, so that

$$12 = 3 \cdot 2 \cdot 2 = 2 \cdot 3 \cdot 2 = 2 \cdot 2 \cdot 3$$

It does not mean that the prime factors are unique (although this is a common misconception), since

$$12 = 3 \cdot 2 \cdot 2 = (-3) \cdot (-2) \cdot 2$$

You may feel that replacing 3 and 2 by their negatives is artificial, but you cannot argue that 3 equals -3, although, come to think of it, they are associates. At least it should be clear that if we are given a factorization, we can always change it by inserting units and their inverses. This is annoying.

You probably know the way out of this morass, at least in the case of the integers. You factor an integer into *positive* primes and then prefix a minus sign if need be. This amounts to choosing, say between the associates 3 and -3, the 3 as a representative of the pair of associates 3 and -3.

We describe an analogous technique for a general ring. Let P be the set of primes in the ring R. For each $p \in P$, denote by $a(p)$ the set of associates of p (these are all in P). For $p, q \in P$, the sets $a(p)$ and $a(q)$ are either equal (if p and q are associates) or disjoint (if not). Thus P is partitioned into disjoint sets of associated primes. In the case of an integer prime p, the set $a(p)$ is equal to $\{p, -p\}$. Now choose one prime from each of the disjoint sets of associates,[1] and call D the resulting subset of P. We say that D is a *designated set of primes* for R. You should see that there are many, many ways that D may be chosen. In the case of the integers, for example, there is nothing to prevent our choosing the *negative* primes as designated! The defining property of a designated set of primes is the following: for each prime $p \in R$, there is a unique element of D which is an associate of p.

In the situation of D being the negative integer primes, we would factor

$$12 = -1(-2)^2(-3)$$

Abstractly, the designated prime factors of 12 determine a subset $\{-2, -3\}$ of the set D of designated primes. We factor 12 as the elements of this subset raised to powers, and then we need a unit out in front. Now we can say what

---

[1]To do the choosing in an arbitrary domain, we are using the *Axiom of Choice*, a set-theoretic axiom that is usually thought to be non-elementary. See [**10**]. We will not need to use this Axiom in any formal proof in this course.

uniqueness means, by doing this in general. The proof of the following involves a straightforward induction argument.

THEOREM 3.5. *Let R be a commutative domain, and let D be a set of designated primes for R. Let $n \in R$, and suppose that n factors into primes of R. Then there is a unique unit v of R and a unique function μ mapping D into the non-negative integers such that*

$$n = v \cdot \prod_{p \in D} p^{\mu(p)}$$

This product is finite, since only finitely many $\mu(p)$ are not 0.

Thus we have uniqueness of all ingredients of factorization, *once a designated set of primes has been chosen*, but this designated set is in no way unique.

Because irreducibles are not always prime, the previous theorems say nothing about factorization into irreducibles. As will be illustrated in an exercise below, factorization into irreducibles does not have to be unique, even up to associates. More importantly, notice that it is a *hypothesis* each time that the ring element n factors into primes. We are not asserting that elements always factor into primes; in fact, such a factorization does not exist in general! We invent a definition to identify the hypothesis of factorization into primes. We say that the commutative domain R is a *unique factorization domain (UFD)* if every $n \in R$ which is not a unit and not zero can be factored into finitely many primes. The content of the Fundamental Theorem of Arithmetic is that the integers is a UFD.

2. Unique Factorization

The goal of this section is to prove two theorems that identify UFD's. These theorems will be used throughout the rest of the course. The first theorem says that if a ring has a kind of "division theorem," then all its ideals are principal ideals. The second theorem says that if all the ideals of a domain are principal, then the ring is a UFD.

We write $\mathbb{N}$ for the non-negative integers. Let R be a domain. A *Euclidean norm* on R is a function $N : R \to \mathbb{N}$ such that

(a) $N(0) = 0$.

(b) If $n, m \in R$ with $n \neq 0$, then there are $q, r \in R$ such that $m = q \cdot n + r$ and $N(r) < N(n)$.

We claim that the absolute value function is a Euclidean norm for the integers. Indeed, $|0| = 0$, and if $m, n \in \mathbb{Z}$ with $n \neq 0$, then apply the Division Theorem to divide m by $|n|$. There are $q, r \in \mathbb{Z}$ such that $m = q \cdot |n| + r$ and $0 \leq r < |n|$. In other words, $|r| < |n|$. That was easy.

In two of our subsequent chapters, we will give other examples of Euclidean norms. For this chapter, we want to keep focus on proving our two main theorems. Here is the first one, generalizing Theorem 2.10.

THEOREM 3.6. *Let R be a domain with a Euclidean norm. Then every ideal of R is a principal ideal.*

PROOF. Let N be a Euclidean norm for R. Let I be an ideal of R. If I consists only of 0, then $I = 0R$ is principal. Otherwise, since N maps to the non-negative integers, well-ordering finds $n \in I$ such that $N(n)$ is minimal among $N(x)$ for all $x \in I$ with $x \neq 0$. We will show that $I = nR$. Indeed, the definition of ideal shows that $nR \subseteq I$.

We have $n \neq 0$. Let $m \in I$, and the property of the norm finds $q, r \in R$ such that $m = q \cdot n + r$ and $N(r) < N(n)$. Since $m, n \in I$, we see that $m - qn \in I$. This says that $r \in I$. The minimality of $N(n)$ shows that r cannot be non-zero. Thus, $r = 0$, and so $m = qn$, so that $m \in nR$, as needed. $\square$

A domain in which all ideals are principal is called a *principal ideal domain*, abbreviated to *PID*. Our second main theorem will be that every PID is a UFD.

To begin working toward this theorem, we recall that primes are always irreducible in a domain but that irreducibles are not always prime. In a PID, irreducibles are prime, just as in the integers.

THEOREM 3.7. *If R is a PID, and if p is an irreducible in R, then p is prime.*

PROOF. Let p be an irreducible, and suppose that it divides the product ab of elements of R. We use a clever idea that goes back in some form to Euclid.[2] Define I to be the set of $ar + ps$ where $r, s \in R$. We see that $I = aR + pR$. An exercise on p.49 showed that the sum of ideals is an ideal; therefore, I is an ideal of R. By hypothesis, I is principal, and so it can be written tR for some $t \in R$.

Observe that $I = tR$ contains pR, and so t divides p. Since p is irreducible we must have either t is a unit or t is an associate of p.

If t is a unit, then by Proposition 3.2, we have that $tR = R$. In particular, we see that $1 \in I$, and we can write $1 = ar + ps$ for some $r, s \in R$. Then

$$b = bar + bps$$

and since p divides ab and p divides ps we see that p divides b, as needed.

If t is an associate of p, then Proposition 3.2 forces that $I = tR = pR$. In this case, the element $a = a \cdot 1 + p \cdot 0$ of I must be a multiple of p, thus p divides a.   □

Now we must work toward showing that every element of a PID factors into irreducibles (which are primes!). This step requires a substitute for integer induction in a general PID – the Ascending Chain Lemma. This Lemma

---

[2]See [**7**, Book X, Proposition 3].

goes back to E. Noether, who was one of the principle pioneers of general ring theory.[3] The ideals I_k in the statement of the Lemma are called an *ascending chain*. One way to state the Lemma is to say, "in a PID, an ascending chain must eventually repeat." Rings for which this lemma is valid are called *noetherian rings*. Many rings which are not PID's are nonetheless noetherian, and the study of these rings is rich.

ASCENDING CHAIN LEMMA. *Let R be a PID, and let I_k be ideals of R for $k = 1, 2, 3, \ldots$. Suppose that $I_k \subseteq I_{k+1}$ for all k. Then there is a positive integer n such that $I_n = I_{n+1}$.*

PROOF. Let I be the union of all the I_k, and it is an exercise to show that I is an ideal of R. (This uses $I_k \subseteq I_{k+1}$, and it is not true in general that a union of ideals is an ideal.)

Because R is a PID, we can write $I = aR$ for some $a \in R$. Now a, being an element of I, is an element of I_n for some n. We claim that I_n is equal to I_{n+1}.

Indeed, the hypothesis shows that I_{n+1} contains I_n. But also, $a \in I_n$ implies

$$aR \subseteq I_n \subseteq I_{n+1} \subseteq I = aR$$

shows that all these ideals must be equal. This proves the lemma.   □

It may help to describe a simpler proof of the Ascending Chain Lemma for the integers. The ideals I_k look like $a_k \mathbb{Z}$ for positive integers a_k. Then the statement "$I_{k+1} \supseteq I_k$" may be translated "a_{k+1} divides a_k." In other words, a_1 is divisible by a_2 is divisible by a_3, and so on; note that the absolute values of $a_1, a_2, \ldots$ must be non-increasing. Well-ordering in the integers then says that there must be a_n of minimum absolute value. Since a_{n+1} divides a_n, this minimality forces that a_n and a_{n+1} are associates, hence $I_n = I_{n+1}$. Our use of

---

[3]Noether's contributions are detailed in the many references to her work in [**6**].

well-ordering (equivalent to induction) shows why we refer to the Ascending Chain Lemma as a substitute for induction.

Now we get some factorization.

THEOREM 3.8. *Let R be a noetherian domain, and let $x \in R$ be neither zero nor a unit. Then x factors as a product of finitely many irreducibles.*

PROOF. Assuming that the theorem is false, let S be the set of elements of R which are not zero, not units, and do not factor as a product of irreducibles. Choose $x \in S$.

The goal of the following is to produce a sequence a_k of elements of S with the property that $a_{k-1} = a_k \cdot b_k$ where $a_k \in S$ and b_k is not a unit.

We get started by setting $a_0 = x$. Let us show how to get a_1. The element x cannot itself be irreducible, and so it factors $x = a \cdot b$ where $a, b \in R$ are neither units. We claim that one of a or b is an element of S. We have remarked that they are not units, and neither can be zero, since x is not zero. If a and b both factored into irreducibles, then x would be the product of irreducibles, contrary to $x \in S$. Thus a or b is in S, say $a \in S$. If we put $a_1 = a$ and $b_1 = b$ then we have $a_0 = a_1 \cdot b_1$ as needed.

Now assume we have a_k, and we will show how to get the equation for $k = j + 1$. We know from $a_k \in S$ that Then a_k is not irreducible, and so it can be factored $a_k = c \cdot d$ where $c, d \in R$ are neither units. Exactly as with x, it follows that one of c or d is in S, say $c \in S$. Put $a_{k+1} = c$ and $b_{k+1} = d$, and we continue the pattern. We have the desired equation for all positive integers k.

We see that a_k divides a_{k-1} and so $a_{k-1}R \subseteq a_k R$ for all $k \geq 1$. Since R is noetherian, there is some positive integer n such that $a_n R = a_{n+1}R$. Proposition 3.2 then says that $a_n = v a_{n+1}$ for some unit v. We also have $a_n = a_{n+1} \cdot b_{n+1}$. Since a_{n+1} is not zero and R is a domain, we have that $b_{n+1} = v$, which contradicts that b_{n+1} is not a unit. $\square$

We have only left to assemble the pieces for our second main theorem.

THEOREM 3.9. *Let R be a PID. Then R is a UFD.*

PROOF. Let $n \in R$ with n not zero and not a unit. By the Ascending Chain Lemma, R is noetherian, and so Theorem 3.8 shows that n factors into a product of finitely many irreducibles. By Theorem 3.7, these irreducibles are primes, and so n is a product of finitely many primes, as needed.   □

Except for the last sentence of the proof, the only property of the ring R which was used was that it is noetherian. Have we proved that noetherian domains are UFD's? We certainly have proved that elements of noetherian domains factor into irreducibles. The trouble is that irreducibles need not be primes. Later we will be able to give an example of a noetherian ring which is not a UFD.

There is one further topic to cover: greatest common divisor. The definition for the integers can easily be generalized. Let R be a domain. Given $a, b \in R$ we say that $c \in R$ is a *greatest common divisor* for a, b if

(1) c divides a, and c divides b;
(2) if $d \in R$ divides both a and b, then d divides c.

The phrase *greatest common divisor* will be abbreviated *GCD*. If c is a GCD of a, b, then so is every associate of c. Thus, GCD's are not necessarily unique, and we insist on the indefinite article: *a* GCD. Nonetheless, for PID's we can prove a result very similar to that for the integers.

THEOREM 3.10. *Let R be a PID, and let $a, b \in R$ not both zero. Then*

(a) there is $c \in R$ which is a GCD for a and b.
(b) $c = xa + yb$ for some $x, y \in R$.
(c) An element of R is a GCD for a, b if and only if it is an associate of c.

PROOF. Let $I = aR + bR$, so that I is an ideal of R. Write $I = cR$, and you are left the task of providing the rest of the proof!   □

In a later chapter, we will need the following. It generalizes Proposition 2.17.

COROLLARY 3.11. *Let R be a PID. If $p \in R$, then the ideal pR is maximal if and only if p is prime. If p is prime and if $x \in R$ is not divisible by p, then there are $r, s \in R$ such that $rp + sx = 1$.*

PROOF. Let p be prime. Since p is not a unit, the ideal pR is proper. If aR is an ideal containing pR, then a divides p. Since p is irreducible, either a is an associate of p (and $aR = pR$) or a is a unit (and $aR = R$). This shows that pR is maximal.

Conversely, if I is maximal, then write $I = pR$ and p is not a unit. If a divides p, then aR contains pR, and so since pR is maximal, we have either $aR = pR$ or $aR = R$. In other words, either a is an associate of p or a is a unit. This proves that p is irreducible, hence prime.

Suppose that p is prime, $x \in R$, and p does not divide x. Theorem 3.10 finds a GCD for p, x, call it c. Could c be an associate of p? If so, then the hypothesis that p does not divide x is contradicted by the fact that c divides x. Since c is not an associate of p, it must be that c is a unit, so that by Theorem 3.10, 1 is a GCD of p and r. That theorem write $rp + sx = 1$ as required.   □

If R is a PID and if $p \in R$ is prime, Proposition 2.16 now tells us that R/pR is a field.

3. Problems

1. Let S be a subset of the positive integer primes. Define $\mathbb{Q}_S$ to be the set of rational numbers that can be written x/y where x and y are integers, and if p is a positive integer prime divisor of y, then $p \in S$. Show that $\mathbb{Q}_S$ is a subring of $\mathbb{Q}$. (Note: it is possible for S to be empty or finite or infinite.)

2. Continuing the notation of the previous problem, what are the units of $\mathbb{Q}_S$? What are the primes?

3. Let R be a subring of $\mathbb{Q}$.

(a) Show that $\mathbb{Z} \subseteq R$.

(b) Define S to be the set of positive integer primes p such that $1/p \in R$. Show that $\mathbb{Q}_S \subseteq R$.

(c) Let $x/y \in R$ with $x, y \in \mathbb{Z}$ in lowest terms, and suppose that p is a prime divisor of y. Show that there is an element $z/p \in R$ with $z \in \mathbb{Z}$ not divisible by p.

(d) Suppose that $z/p \in R$ with z not divisible by p. Show that $1/p \in R$, so that $p \in S$. (Hint: GCD of z, p?)

(e) Conclude that $R = \mathbb{Q}_S$.

4. Let S be a subset of the positive integer primes. Show that the ring $\mathbb{Q}_S$ is a PID. (Hint: if I is an ideal of $\mathbb{Q}_S$, show that $I \cap \mathbb{Z}$ is an ideal of $\mathbb{Z}$.)

5. Let R be a commutative domain, and let $p \in R$ not be 0 and not be a unit. Show that p is prime if and only if $R/(pR)$ is a domain. (Recalling a previous problem, we can say this another way: p is prime if and only if pR is a prime ideal.)

6. Let p be an integer prime. Show that $\sqrt{p}$ is a prime in $\mathbb{Z}[\sqrt{p}]$.

7. Show that 3 is prime in $\mathbb{Z}[i]$. Show that 5 is *not* prime in that ring.

8. Show that every associate of a prime is a prime. Show that every associate of an irreducible is an irreducible.

9. Let $\alpha \in \mathbb{C}$ with $\alpha \notin \mathbb{Z}$ and $\alpha^2 \in \mathbb{Z}$. Define $\mathbb{Z}[\alpha]$ as usual: the set of $a + b \cdot \alpha$ where $a, b \in \mathbb{Z}$. We will show that $\mathbb{Z}[\alpha]$ is noetherian.[4] Let I be a non-zero ideal of $\mathbb{Z}[\alpha]$.

(a) Show that there is a positive integer $n \in I$.
(b) Show that every coset of I in $\mathbb{Z}[\alpha]$ can be written $a + b \cdot \alpha + I$ where $a, b \in \mathbb{Z}$ and $0 \leq a, b < n$.
(c) Show that $\mathbb{Z}[\alpha]/I$ has at most n^2 elements.
(d) Show that every ascending chain of ideals of $\mathbb{Z}[\alpha]$ has a repeat.

10. Let I be a proper ideal in a noetherian domain R. Show that I is *finitely generated* – that there are $a_1, \ldots, a_n \in I$ such that every element of I can be written

$$\sum_{j=1}^{n} r_j \cdot a_j \quad \text{for some} \quad r_1, \ldots, r_n \in R$$

(Hint: if not, for $a_1, \ldots, a_n \in I$, choose $a_{n+1} \in I \setminus \sum_{j=1}^{n} R a_j$.)

11. Suppose that R is a domain in which every proper ideal is finitely generated. Show that R is noetherian. (Hint: for an ascending chain, let I be the union of the ideals.)

12. Complete the steps to show that 2 is an irreducible element of $\mathbb{Z}[\sqrt{-5}]$. Recall that the map $f : \mathbb{Z}[\sqrt{-5}] \rightarrow \mathbb{Z}[\sqrt{-5}]$ defined by $f(a + b \cdot \sqrt{-5}) = a - b \cdot \sqrt{-5}$ is a ring isomorphism.

(a) If $\alpha \in \mathbb{Z}[\sqrt{-5}]$, and if $\alpha \cdot f(\alpha) < 5$, then $\alpha \in \mathbb{Z}$.
(b) If $2 = \alpha \cdot \beta$ in $\mathbb{Z}[\sqrt{-5}]$, then show that $2 \cdot f(2) = \alpha \cdot f(\alpha) \cdot \beta \cdot f(\beta)$.
(c) Show that (a) and the equation of (b) lead to $\alpha = \pm 2$ or $\beta = \pm 2$.

---

[4]We will generalize the argument that $\mathbb{Z}[\sqrt{2}]/I$ is finite, where I is a non-zero ideal.

13. Show that 2 is not a prime element of $\mathbb{Z}[\sqrt{-5}]$, by considering the fact that $6 = (1 + \sqrt{-5}) \cdot (1 - \sqrt{-5})$.

14. Suppose that R is a UFD. Show that every irreducible in R is prime.

15. Show that $\mathbb{Z}[\sqrt{-5}]$ is noetherian but not a UFD. (Hint: use a couple of previous problems.)

16. Let D be a set of designated primes for the UFD R. Let $m, n \in R$ and suppose they are factored like this:

$$m = u \cdot \prod_{p \in D} p^{\mu(p)} \quad \text{and} \quad n = v \cdot \prod_{p \in D} p^{\nu(p)}$$

where u, v are units. For each $p \in D$, define $\pi(p)$ to be the minimum of $\mu(p)$ and $\nu(p)$. Show that the following is a GCD of m, n.

$$\prod_{p \in D} p^{\pi(p)}$$

(One detail: make sure you explain why all but finitely many of the $\pi(p)$ are 0; not difficult but important.)

17. Let R be a UFD. Define a *least common multiple* for arbitrary $m, n \in R$, and prove that a least common multiple always exists. (Note: do not assume that elements of R can be ordered; *least* cannot be defined as *smallest*.)

18. Let R be a PID and $m, n \in R$. Show that c is a least common multiple for m, n if and only if $mR \cap nR = cR$. (Note: this approach cannot be used in the UFD problem, since the UFD might not be a PID.)

CHAPTER 4

Applications

Diophantine problems seek integer solutions to algebraic equations. There are many such problems; some of the most famous took centuries to solve.[1] We will identify several number rings that are PID's, and use this information to attack some Diophantine problems. For instance, we will prove Fermat's Last Theorem in the case of exponent 3 and 4!

1. Some Quadratic PID's

If r, s are integers, then the roots of the polynomial $X^2 + r \cdot X + s$ involve $\sqrt{r^2 - 4s}$. Define $m = r^2 - 4s$, and suppose that $\sqrt{m}$ is irrational.[2]

We know some things about the set $\mathbb{Q}[\sqrt{m}]$:

(a) Each element of $\mathbb{Q}[\sqrt{m}]$ can be written uniquely as $a + b \cdot \sqrt{m}$, where $a, b \in \mathbb{Q}$.

(b) The set $\mathbb{Q}[\sqrt{m}]$ is a field.

(c) The function $\sigma : \mathbb{Q}[\sqrt{m}] \to \mathbb{Q}[\sqrt{m}]$ defined by $\sigma(a + b \cdot \sqrt{m}) = a - b \cdot \sqrt{m}$ is a ring isomorphism.

For some particular values of m, the ring $Q[\sqrt{m}]$ has a subring that is a PID – we will be able to describe the units and primes. To build them, we

---

[1] The long history of Diophantine problems is detailed up to the early 1900's in the three volumes of [**3**]. See also [**2**] for many examples.

[2] If $m < 0$, there are two complex numbers whose square is m; choose either one to use as $\sqrt{m}$.

give the names δ and δ' to the two roots of $X^2 + rX + s$:

$$\delta = \frac{-r + \sqrt{m}}{2} \quad \text{and} \quad \delta' = \frac{-r - \sqrt{m}}{2}$$

The factorization

$$X^2 + r \cdot X + s = (X - \delta) \cdot (X - \delta')$$

shows that

(4.1) $\qquad\qquad\qquad \delta + \delta' = -r \quad \text{and} \quad \delta \cdot \delta' = s$

Notice that δ and δ' are elements of $\mathbb{Q}[\sqrt{m}]$. Because r, s are integers, it is an exercise to prove that the set $\mathbb{Z}[\delta]$, consisting of $a + b \cdot \delta$ for all integers a, b, is a subring of $\mathbb{Q}[\sqrt{m}]$. Notice also that $\delta' = -r - \delta$ is an element of $\mathbb{Z}[\delta]$.

Recall the isomorphism σ. We compute that $\sigma(\delta) = \delta'$, using the definitions:

$$\sigma(\delta) = \sigma \left(-\frac{1}{2} \cdot r + \frac{\sqrt{m}}{2} \right)$$

$$= -\frac{1}{2} \cdot r - \frac{\sqrt{m}}{2}$$

$$= \delta'$$

Since σ is additive and maps each rational number to itself, if $a, b \in \mathbb{Z}$, then we see that

$$\sigma(a + b \cdot \delta) = a + b \cdot \sigma(\delta) = a + b \cdot \delta'$$

This shows that σ maps $\mathbb{Z}[\delta]$ to itself, and it is easy to see that this mapping is onto. Thus, σ is a ring isomorphism from $\mathbb{Z}[\delta]$ to itself.

Define $N : \mathbb{Q}[\sqrt{m}] \to \mathbb{Q}$ by $N(\alpha) = \alpha \cdot \sigma(\alpha)$. This important function is called the *algebraic norm*. We have

$$N(\alpha \cdot \beta) = \alpha \cdot \beta \cdot \sigma(\alpha \cdot \beta) = \alpha \cdot \beta \cdot \sigma(\alpha) \cdot \sigma(\beta)$$

$$= \alpha \cdot \sigma(\alpha) \cdot \beta \cdot \sigma(\beta) = N(\alpha) \cdot N(\beta)$$

The definition also shows that if $\alpha \neq 0$, then $N(\alpha) \neq 0$. For $a, b \in \mathbb{Q}$, the definition of σ and the equation (4.1) allow us to compute $N(a + b \cdot \delta)$ in terms of a, b.

$$
\begin{aligned}
N(a + b \cdot \delta) &= (a + b \cdot \delta) \cdot (a + b \cdot \delta') \\
&= a^2 + a \cdot b \cdot (\delta + \delta') + b^2 \cdot \delta \cdot \delta' \\
&= a^2 - a \cdot b \cdot r + b^2 \cdot s
\end{aligned}
$$

This formula shows that N maps $\mathbb{Z}[\delta]$ into the *integers*. Furthermore, if $\alpha, \beta \in \mathbb{Z}[\delta]$ and α divides β in $\mathbb{Z}[\delta]$, then $N(\alpha)$ divides $N(\beta)$ in $\mathbb{Z}$.

We will deal with specific examples where the hypothesis of the following proposition holds. A paraphrase of this hypothesis: you can approximate elements of $\mathbb{Q}[\sqrt{m}]$ by elements of $\mathbb{Z}[\delta]$.

PROPOSITION 4.1. *Assume the notation just given for $\mathbb{Z}[\delta]$. Suppose that for each $\alpha \in \mathbb{Q}[\sqrt{m}]$, there is $\gamma \in \mathbb{Z}[\delta]$ such that $|N(\alpha - \gamma)| < 1$. Then $|N|$ is a Euclidean norm on $\mathbb{Z}[\delta]$, so that this ring is a PID.*

PROOF. Notice that $N(0) = 0$, and so we need to establish the quotient/remainder formula.

Let $\alpha, \beta \in \mathbb{Z}[\delta]$ with $\beta \neq 0$. Then α and β are elements of the *field* $\mathbb{Q}[\sqrt{m}]$, and so α/β is in that field, too. By the hypothesis, there is $\gamma \in \mathbb{Z}[\delta]$ such that $|N(\alpha/\beta - \gamma)| < 1$. Multiplying by $N(\beta)$, we have

$$
|N(\beta) \cdot N(\alpha/\beta - \gamma)| < |N(\beta)|
$$

so that

$$
|N(\alpha - \gamma \cdot \beta)| < |N(\beta)|
$$

This is precisely the quotient/remainder result needed. $\square$

Under the assumption that $\mathbb{Z}[\delta]$ is a PID, we examine the units and primes. We claim that z is a unit in $\mathbb{Z}[\delta]$ if and only if $N(z) = \pm 1$. Indeed, if z is a unit, write $z \cdot w = 1$, and then $N(z) \cdot N(w) = N(1) = 1$. Since $N(z), N(w)$ are integers, we see that $N(z) = \pm 1$. Conversely, if $N(z) = \pm 1$, then since $N(z) = z \cdot \sigma(z)$, we see that z divides 1.

For an integer n, we have $N(n) = n^2$, and so the only integer units in $\mathbb{Z}[\delta]$ are the integers ± 1. The set $\mathbb{Z}[\delta]$ can have finitely many or infinitely many non-integer units.

Next we claim that each prime in $\mathbb{Z}[\delta]$ divides some integer prime. For if π is prime in $\mathbb{Z}[\delta]$, then $N(\pi)$ is not 0 and it is not ± 1, and so it factors into integer primes. Then $N(\pi) = \pi \cdot \sigma(\pi)$ shows that π must divide at least one of the integer prime factors of $N(\pi)$, and this shows that π divides an integer prime, as claimed. This shows that we can collect the primes in $\mathbb{Z}[\delta]$ by considering the factorization of each integer prime.

Let p be an integer prime, and suppose that π is a prime in $\mathbb{Z}[\delta]$ dividing p. Then $N(\pi)$ divides $N(p) = p^2$. Since the integer $N(\pi)$ cannot be ± 1, we have $N(\pi) = \pm p$ or $N(\pi) = \pm p^2$. (By the way, this proves that each prime in $\mathbb{Z}[\delta]$ divides a *unique* positive integer prime.)

For a given integer prime p, let us see that there are three possible types of factorization in $\mathbb{Z}[\delta]$. If $N(\pi) = \pm p^2$, we claim that p is prime in $\mathbb{Z}[\delta]$ (so that p is an associate of π). Indeed, write $p = \pi \cdot \alpha$, and then $p^2 = N(p) = N(\pi) \cdot N(\alpha)$, so that since $N(\pi) = \pm p^2$, we see that $N(\alpha) = \pm 1$. Thus, α is a unit, and p is an associate of π. In this case, we say that the prime p *extends* to $\mathbb{Z}[\delta]$. We have proved that p extends if $N(\pi) = \pm p^2$. (We will eventually prove the converse.)

Suppose p is not prime in $\mathbb{Z}[\delta]$, and then $N(\pi) = \pm p$. We have $\pi \cdot \sigma(\pi) = \pm p$. Also, $\sigma(\pi)$ is prime. If π and $\sigma(\pi)$ are not associates, then we say that p *splits* in $\mathbb{Z}[\delta]$. If π and $\sigma(\pi)$ are associates, we say that p *ramifies* in $\mathbb{Z}[\delta]$.

Thus, integer primes either extend, split, or ramify in $\mathbb{Z}[\delta]$. Determining which are which comes down to the factorization of $X^2 + r \cdot X + s$ in $\mathbb{Z}_p$.

Claim. Let p be a positive integer prime.

(a) If $X^2 + r \cdot X + s$ has no roots in $\mathbb{Z}_p$, then p extends.

(b) If $X^2 + r \cdot X + s$ has two distinct roots in $\mathbb{Z}_p$, then p splits.

(c) If $X^2 + r \cdot X + s$ has exactly one root in $\mathbb{Z}_p$, then p ramifies.

PROOF. First, let π be a prime in $\mathbb{Z}[\delta]$ that divides p, and suppose that $N(\pi) = \pm p$. We claim that $X^2 + rX + s$ has a root mod p. Write $\pi = a + b \cdot \delta$, and we claim that p does not divide b. Taking the formula for N, we see that

$$\pm p = N(\pi) = a^2 - r \cdot a \cdot b + s \cdot b^2$$

If p divides b, this equation shows that p divides a. It follows that p^2 divides a^2, and it divides $a \cdot b$, and it divides b^2. This implies that p^2 divides p, a contradiction. We conclude that p does not divide b. Since p does not divide b, there is an integer c such that $-a \equiv c \cdot b$. Direct calculation using the formula for $N(\pi)$ will show that c is a root of $X^2 + r \cdot X + s$ in $\mathbb{Z}_p$.

$$0 \equiv \pm p = a^2 - r \cdot a \cdot b + s \cdot b^2$$
$$\equiv c^2 \cdot b^2 + r \cdot c \cdot b^2 + s \cdot b^2$$
$$= b^2 \cdot (c^2 + r \cdot c + s)$$

Since p does not divide b, we see that $c^2 + r \cdot c + s \equiv 0$, so that c is a root in $\mathbb{Z}_p$, as claimed.

The contrapositive of our first claim: if $X^2 + rX + s$ has no roots mod p, then $N(\pi) \neq \pm p$. We conclude that $N(\pi) = \pm p^2$, whence p extends, as shown above. This proves statement (a).

We prove the converse of (a): if p is prime in $\mathbb{Z}[\delta]$, then $X^2 + rX + s$ has no roots mod p. Indeed, if $c \in \mathbb{Z}$ is a root of $X^2 + rX + s$ mod p, then the integer $c^2 + rc + s$ is divisible by p. Notice that

$$c^2 + rc + s = (c - \delta) \cdot (c - \delta')$$

Since p is prime in $\mathbb{Z}[\delta]$, it must be that p divides $c - \delta$ or p divides $c - \delta' = c + r + \delta$. It follows that p divides 1 in the integers – a contradiction. We see that c cannot exist.

For (b) and (c), since $X^2 + r \cdot X + s$ has a root in $\mathbb{Z}_p$, the converse of (a) shows that p is not prime in $\mathbb{Z}[\delta]$. Let π be a prime in $\mathbb{Z}[\delta]$ that divides p, and we know that $N(\pi) = \pm p$. If c is a root of $X^2 + rX + s$ mod p, then, p divides $c^2 + rc + s$, and so π divides $c^2 + rc + s$. As before, we can factor $c^2 + rc + s = (c - \delta) \cdot (c - \delta')$. This proves that π divides $c - \delta$ or π divides $c - \delta'$. In the latter case, $\sigma(\pi)$ divides $c - \delta$; replace π by $\sigma(\pi)$. Now we have π divides $c - \delta$ and $\sigma(\pi)$ divides $c - \delta'$.

Suppose that c is a repeated root of $X^2 + rX + s$ mod p. Then

$$X^2 + r \cdot X + s \equiv (X - c)^2 = X^2 - 2 \cdot c \cdot X + c^2$$

and then $-2c \equiv r$ mod p, so that p divides $2c + r$. Also, π divides $2c + r$, and so π divides $(2c + r) - (c - \delta) = c + r + \delta = c - \delta'$. Both π and $\sigma(\pi)$ divide $c - \delta'$; if they are not associates, then their product divides $c - \delta'$. Thus, $p = \pm N(\pi)$ divides $c - \delta'$. This is impossible as before. We conclude that π and $\sigma(\pi)$ are associates, and p ramifies.

Suppose that p ramifies, so that π and $\sigma(\pi)$ are associates. As above, $\sigma(\pi)$ divides $c - \delta'$, and so π divides $c - \delta'$. Thus, π divides both $c - \delta$ and $c - \delta'$. It follows that π divides $(c - \delta) + (c - \delta') = 2c + r$. Since $2c + r$ is divisible by a prime, it is not a unit. It follows that it factors *in the integers* into primes, and one of those primes is divisible by π in $\mathbb{Z}[\delta]$. Since p is the unique positive integer prime divisible by π, we conclude that p divides $2c + r$, and so $r \equiv -2c$. Compute

$$0 \equiv c^2 + rc + s \equiv c^2 + (-2c) \cdot c + s = -c^2 + s$$

This shows that $s \equiv c^2$ mod p. Thus,

$$X^2 + rX + s \equiv X^2 - 2cX + c^2 = (X - c)^2$$

and c is a repeated root of $X^2 + rX + s$.

For (b), we know that p does not extend and it does not ramify. Now $N(\pi) = \pm p$ and $N(\pi) = \pi \cdot \sigma(\pi)$, where π and $\sigma(\pi)$ are not associates. Thus, p splits. $\qquad\square$

Now that we know how to find the units and primes in rings that satisfy the hypothesis of Proposition 4.1, we give some specific examples.

Claim. If $m = -2, -1, 2, 3$, then the hypothesis of Proposition 4.1 is satisfied with $\delta = \sqrt{m}$ and polynomial $X^2 - m$ on the ring $\mathbb{Z}[\sqrt{m}]$.

PROOF. Let $\alpha \in \mathbb{Q}[\sqrt{m}]$, and we must find $\gamma \in \mathbb{Z}[\delta]$ with $|N(\alpha - \gamma)| < 1$. We can write $\alpha = a + b \cdot \delta$, and since a, b are rational numbers, there are integers x, y such that $|a - x| \le 1/2$ and $|b - y| \le 1/2$. Let $\gamma = x + y \cdot \delta$, so that $\alpha - \gamma = (a - x) + (b - y) \cdot \delta$. Write $c = a - x$ and $d = b - y$, and we have that c, d are rational and that $|c| \le 1/2$ and $|d| \le 1/2$. We need to show that $|N(c + d \cdot \delta)| < 1$.

We have
$$|N(c + d \cdot \sqrt{m})| = |c^2 - m \cdot d^2|$$
In the case that $m = -2, -1$, we have
$$|c^2 - m \cdot d^2| \le \frac{1}{4} + 2 \cdot \frac{1}{4} < 1$$
as needed.

In the case that $m = 2, 3$, we compute

(4.2) $$c^2 - m \cdot d^2 \le c^2 \le \frac{1}{4}$$

and

(4.3) $$m \cdot d^2 - c^2 \le m \cdot d^2 \le 3 \cdot \frac{1}{4}$$

The inequalities (4.2) and (4.3) taken together prove that $|c^2 - m \cdot d^2| < 1$, as needed. $\qquad\square$

Suppose that m is an integer that is not an integer square and with $m \equiv$ 1 mod 4. Define

$$\delta = \frac{1 + \sqrt{m}}{2}$$

and observe that δ is a root of the polynomial $X^2 - X + (1 - m)/4$. Since $m \equiv 1$ mod 4, this polynomial has *integer* coefficients.

Claim. If $m = -11, -7, 5, 13$, then the hypothesis of Proposition 4.1 is satisfied with $\delta = (1 + \sqrt{m})/2$ and polynomial $X^2 - X + (1 - m)/4$ in the ring $\mathbb{Z}[\delta]$.

PROOF. Let $\alpha \in \mathbb{Q}[\sqrt{m}]$; as above, write $\alpha = a + b \cdot \delta$. Choose an integer y such that $|b - y| \leq 1/2$. Now choose an integer x so that

$$\left| a + \frac{b - y}{2} - x \right| \leq \frac{1}{2}$$

Put $\gamma = x + y \cdot \delta$, and we claim that $|N(\alpha - \gamma)| < 1$. We have $\alpha - \gamma = (a - x) + (b - y) \cdot \delta$.

We need to rewrite N, starting with the formula in terms of $a - x$ and $b - y$.

$$N((a - x) + (b - y) \cdot \delta) = (a - x)^2 + (a - x) \cdot (b - y) + \frac{1 - m}{4} \cdot (b - y)^2$$

$$= (a - x)^2 + (a - x) \cdot (b - y) + \frac{(b - y)^2}{4} - m \cdot \frac{(b - y)^2}{4}$$

$$= \left(a - x + \frac{b - y}{2} \right)^2 - m \cdot \left(\frac{b - y}{2} \right)^2$$

We remind you that

$$|a - x| \leq \frac{1}{2} \quad \text{and} \quad \left| a - x - \frac{b - y}{2} \right| \leq \frac{1}{2}$$

If $m = -11, -7$, then

$$\left| \left(a - x + \frac{b - y}{2} \right)^2 - m \cdot \left(\frac{b - y}{2} \right)^2 \right| \leq \left(\frac{1}{2} \right)^2 + 11 \cdot \left(\frac{1/2}{2} \right)^2 = \frac{15}{16}$$

as needed.

If $m = 5, 13$, then

$$\left(a - x + \frac{b-y}{2}\right)^2 - m \cdot \left(\frac{b-y}{2}\right)^2 \leq \left(a - x + \frac{b-y}{2}\right)^2 \leq \frac{1}{4}$$

and

$$m \cdot \left(\frac{b-y}{2}\right)^2 - \left(a - x + \frac{b-y}{2}\right)^2 \leq 13 \cdot \frac{1}{16} < 1$$

and the Claim holds.   □

We determine the units of one of these rings. In particular, our example has infinitely many units.

PROPOSITION 4.2. *Every unit of $\mathbb{Z}[\sqrt{2}]$ has the form $s \cdot (1 + \sqrt{2})^n$, where $s = \pm 1$ and n is an integer.*

PROOF. Notice that $(1 + \sqrt{2}) \cdot (\sqrt{2} - 1) = 1$, so that $1 + \sqrt{2}$ is a unit. In particular, $(1 + \sqrt{2})^n$ is an element of $\mathbb{Z}[\sqrt{2}]$ for all integers n.

Let $a + b \cdot \sqrt{2}$ be a unit in $\mathbb{Z}[\sqrt{2}]$. Then $N(a + b \cdot \sqrt{2}) = a^2 - 2 \cdot b^2 = \pm 1$. Assume first that $a > 0$ and the norm is 1, and we will prove that $a + b \cdot \sqrt{2}$ has the required form by induction on a. We will use the equation $a^2 = 1 + 2b^2$ repeatedly.

Assume that $b = 0$, so that $a^2 = 1$ implies that $a = 1$. The unit $1 = a + b \cdot \sqrt{2}$ has the required form. Assume that $b \neq 0$. We claim that

(4.4) $$0 < 3 \cdot a - 4 \cdot |b| < a$$

Indeed,

$$a^2 = 1 + 2 \cdot b^2 > 2 \cdot b^2 > \frac{16}{9} \cdot b^2$$

Thus, $a^2 > 16 \cdot b^2 / 9$, and so $a > 4 \cdot |b|/3$, and so $3 \cdot a - 4 \cdot |b| > 0$. Also, remembering that $b \neq 0$, we see that

$$4 \cdot a^2 = 4 + 8 \cdot b^2 < 4 \cdot b^2 + 8 \cdot b^2 < 16 \cdot b^2$$

It follows that $2 \cdot a < 4 \cdot |b|$, and so $3 \cdot a < 4 \cdot |b| + a$, and this give the other inequality in (4.4).

Compute that

$$(a + b \cdot \sqrt{2}) \cdot (1 + \sqrt{2})^2 = (3a + 4b) + (3b + 2a) \cdot \sqrt{2}$$
$$\text{and} \quad (a + b \cdot \sqrt{2}) \cdot (1 + \sqrt{2})^{-2} = (3a - 4b) + (3b - 2a) \cdot \sqrt{2}$$

The inequalities (4.4) show that one of the terms on the right has the form $c + d \cdot \sqrt{2}$ with $0 < c < a$. By induction $c + d \cdot \sqrt{2} = (1 + \sqrt{2})^n$ for some integer n, and we see that $a + b \cdot \sqrt{2}$ has the required form. This completes the case $0 < a$ and $N(a + b \cdot \sqrt{2}) = 1$.

Now assume that $a < 0$ and $N(a + b \cdot \sqrt{2}) = 1$. Then $-a - b \cdot \sqrt{2}$ is a unit of norm 1, and $-a > 0$, so that

$$a + b \cdot \sqrt{2} = (-1) \cdot (1 + \sqrt{2})^n \quad \text{for some} \quad n \in \mathbb{Z}$$

Finally, assume that α is a unit in $\mathbb{Z}[\sqrt{2}]$ has norm -1. Then $\alpha \cdot (1 + \sqrt{2})$ has norm 1, and so it has the required form. Then α has the required form, as well. $\qquad\qquad\qquad\qquad\qquad\qquad\qquad\qquad\qquad\qquad\qquad\qquad\square$

Another specific example: the ring $\mathbb{Z}[i]$ is called the *Gaussian integers*. Our polynomial is $X^2 + 1$ and it is easy to apply what we have done and categorize the units and primes in $\mathbb{Z}[i]$. For units, let $N(a + b \cdot i) = \pm 1$, for integers a, b. Since $N(a + b \cdot i) = a^2 + b^2$, we see that the units are the four numbers $\pm 1, \pm i$.

Let p be a positive integer prime, and suppose that $X^2 + 1$ has a root c mod p. Then $c^2 + 1 \equiv 0$ mod p. We see that $-c$ is also a root of this polynomial. If $-c \equiv c$, then $X^2 + 1$ has a repeated root. We have that $-c \equiv c$ implies $0 \equiv 2c$, and so either $0 \equiv 2$ or $0 \equiv c$. The latter possibility is inconsistent with $c^2 + 1 \equiv 0$. Thus, $0 \equiv 2$, so that $p = 2$. Thus, if p ramifies, then $p = 2$. And, in fact, $2 = -i \cdot (1 + i)^2$, and we see that *the prime 2 ramifies*. Also, if $X^2 + 1$

has a root mod p for an odd prime p, then the polynomial has two distinct roots, so that p splits. If $X^2 + 1$ has no root mod p, then p extends.

We claim that p splits if and only if 4 divides $p - 1$. Let p split, so that there are two roots $\pm c$ mod p for $X^2 + 1$. Since $c^2 + 1 \equiv 0$, we see that $c^2 \equiv -1$, and it follows that $c^4 \equiv 1$. This proves that c has order dividing 4. We will show that the order of c is not 1 and not 2, and it will follow that its order is 4. Indeed, if $c^2 \equiv 1$, then $0 \equiv c^2 + 1 \equiv 2$, contradicting the fact that p is odd. Thus, c has order 4. The multiplicative group of the non-zero elements of $\mathbb{Z}_p$ is cyclic of order $p - 1$; it has an element of order 4 if and only if 4 divides $p - 1$. We have proved that if p splits, then 4 divides $p - 1$.

For the converse, suppose that 4 divides $p - 1$. Then $\mathbb{Z}_p$ has an element c having multiplicative order 4. Since $c^4 \equiv 1$, we see that $(c^2 - 1)(c^2 + 1) \equiv 0$. Because c does not have order 2, it is the case that $c^2 - 1 \not\equiv 0$, and so since $\mathbb{Z}_p$ is a field, we get $c^2 + 1 \equiv 0$. Thus, $\pm c$ are distinct roots of $X^2 + 1$, and p splits.

We conclude that *the positive integer prime p splits in $\mathbb{Z}[i]$ if and only if $p - 1$ is divisible by 4.*

We are left with this case: *the positive integer prime p extends in $\mathbb{Z}[i]$ if and only if p is odd and 4 does not divide $p - 1$.*

2. Sums of Two Integer Squares

A Diophantine problem: find all integers n, a, b such that $n = a^2 + b^2$. We could say, "find all integers that can be written as a sum of two squares." We allow a or b to be 0, and so each integer square is a sum of two squares. On the other hand, 13 is not a square, but $13 = 2^2 + 3^2$. A little thought will show that 7 is not a sum of two squares.

This particular problem was solved originally by Fermat. We will use the prime factorization in the Gaussian integers to present a solution.

THEOREM 4.3. *Let n be a positive integer. Let S be the set of positive integer prime factors of n. Then there are integers a and b such that $n = a^2 + b^2$ if and only if for each $p \in S$ such that p is odd and 4 does not divide $p - 1$, the prime p occurs an even number of times in the integer factorization of n.*

PROOF. Assume that n can be a sum of squares *without* satisfying the stated condition. Working by contradiction, suppose $n = a^2 + b^2$ is chosen as small as possible to violate the statement. Then there is a positive odd prime divisor p of n with 4 not dividing $p - 1$ and such that the exponent of p in the prime factorization of n is odd.

In $\mathbb{Z}[i]$ we have p divides $(a + bi)(a - bi)$. We know that p is prime in $\mathbb{Z}[i]$, and so p divides $a + bi$ or p divides $a - bi$. In either case it follows that p divides both a and b in the integers.

Write $a = pc$ and $b = pd$ so that $n = p^2(c^2 + d^2)$ shows that p^2 divides n. We see that n/p^2 is an integer that is a sum of two squares. The minimality of n implies that p occurs an even number of times as a factor of n/p^2. But then p occurs an even number of times in n, and this contradiction completes this half of the proof.

To obtain the other direction of the argument, we pause to prove that if p is either 2 or is odd with 4 dividing $p - 1$, then there are integers a, b such that $p = a^2 + b^2$. Of course, $2 = 1^2 + 1^2$. An odd prime p with 4 dividing $p - 1$ splits in $\mathbb{Z}[i]$. Let $a + b \cdot i$ be a prime divisor of p, and we know that $\pm p = N(a + b \cdot i) = a^2 + b^2$. Since p is positive, we see that $p = a^2 + b^2$ as needed.

Now suppose that n satisfies the factorization condition, and we show that n is a sum of squares. The positive integer prime factors p of n, such that 4 does not divide $p - 1$, occur an even number of times; in other words their product is a square q^2. Notice that q^2 can be written $0^2 + q^2$. The other prime factors p of n are either 2 or have 4 dividing $p - 1$; each of these can be written as a sum of two squares. Thus n is a product of factors which are sums of two squares. We can complete the proof by showing that the product of two integers, each of which is a sum of two squares, is itself a sum of two squares. This last fact is obtained by as follows.

$$
\begin{aligned}
(a^2 + b^2) \cdot (c^2 + d^2) &= N(a + bi) \cdot N(c + di) \\
&= N((a + bi)(c + di)) \\
&= N((ac - bd) + (ad + bc)i) \\
&= (ac - bd)^2 + (ad + bc)^2
\end{aligned}
$$

$\square$

3. Fermat's Last Theorem: Exponent 3

We move to a special case of *Fermat's Last Theorem*, stated by Fermat as a conjecture and finally proved in 1994 by Andrew Wiles[3] – a very significant event in mathematics! Fermat's Last Theorem says that if $n \geq 3$ is an integer, then the equation $x^n + y^n = z^n$ cannot have integer solutions in which x, y, z are all non-zero. Thus, Fermat's Last Theorem presents another Diophantine problem.

We will prove Fermat's Last Theorem in the case $n = 3$. (This case was known long before Wiles.) Define $w = \exp(2\pi i/3)$, so that $w^3 = 1$. Also, $w^2 = -w - 1 = \overline{w}$. The minimal polynomial of w is $X^2 + X + 1$. The ring $\mathbb{Z}[w]$ is equal to the ring $\mathbb{Z}[-w]$; the polynomial for $-w$ is $X^2 - X + 1$, one of those considered in the first section above. We see that $\mathbb{Z}[w]$ is a PID.

---

[3]Wiles combined his own, very striking ideas with an accumulation of theorems developed by many mathematicians over three centuries, a remarkable accomplishment. Wiles' paper on the proof is this: *Modular elliptic curves and Fermat's Last Theorem*, Annals of Mathematics, 1995, Volume 141, Number 3, pp.443-551. An online search will indicate several books discussing and/or explaining this very technical proof.

For integers a, b, we can calculate $N(a+b \cdot w) = a^2 - a \cdot b + b^2$. The element $a + b \cdot w$ is a unit in $\mathbb{Z}[w]$ if and only if $N(a + b \cdot w) = \pm 1$. We see that

$$a^2 - a \cdot b + b^2 = \left(a - \frac{b}{2}\right)^2 + \frac{3}{4} \cdot b^2$$

If $|b| \geq 2$, this cannot be ± 1. If $b = 0$, we see that $a = \pm 1$, giving us the expected units ± 1. If $b = \pm 1$, then we need $|a - b/2| = 1/2$. The solutions are seen to be $a = 1, b = 1$ (unit $1 + w = -w^2 = -\overline{w}$), $a = 0, b = 1$ (unit w), $a = -1, b = -1$ (unit $-1 - w = w^2 = \overline{w}$), $a = 0, b = -1$ (unit $-w$). There are six units.

The integer prime 3 ramifies in $\mathbb{Z}[w]$, since $X^2 + X + 1$ has repeated root 1 in $\mathbb{Z}_3$. Define $\lambda = w - 1$, and we claim that λ is a prime divisor of 3. Indeed, if π is a prime divisor of 3, then π^2 divides 3. Compute $\lambda^2 = w^2 - 2w + 1 = -3w$, and since w is a unit, this shows that π^2 and λ^2 are associates. Since π is prime, we also have that π divides λ. Write $\lambda = \pi \cdot \alpha$, and then $\lambda^2 = \pi^2 \cdot \alpha^2$. Since λ^2 and π^2 are associates, this proves that α is a unit, and now we see that π and λ are associates, and the claim holds.

Next we claim that $\mathbb{Z}[w]/\lambda\mathbb{Z}[w]$ is a field of order 3. It is a field, since $\lambda\mathbb{Z}[w]$ is a maximal ideal. If $a + b \cdot w \in \mathbb{Z}[w]$, then

$$a + b \cdot w = a + b + b \cdot \lambda \quad \text{so that} \quad a + b \cdot w + \lambda\mathbb{Z}[w] = a + b + \lambda\mathbb{Z}[w]$$

The integer $a + b$ can be written $3q + r$ for integers q, r, where $0 \leq r \leq 2$, and $3 \in \lambda\mathbb{Z}[w]$, so that

$$a + b + \lambda\mathbb{Z}[w] = r + \lambda\mathbb{Z}[w]$$

This shows that there are at most 3 elements in the quotient ring. You are left to prove that there cannot be less than 3.

The following lemma does all the (difficult!) technical work needed for the case of Fermat's Theorem we are considering.

LEMMA 4.4. *Let $x, y, z \in \mathbb{Z}[w]$ and let e be a positive integer. Suppose that x, y have no common prime divisor. Suppose also that λ does not divide xyz. Let t, u, v be units in $\mathbb{Z}[w]$. Then $t \cdot x^3 + u \cdot y^3 + v \cdot z^3 \cdot \lambda^{3e} \neq 0$.*

PROOF. Get a counterexample with e as small as possible. We can divide through by t to assume that $x^3 + u \cdot y^3 + v \cdot z^3 \cdot \lambda^{3e} = 0$. We can absorb a minus sign on u into y and assume that $u \in \{1, w, w^2\}$. In any case $u \equiv 1 \bmod \lambda$.

Next we claim that $u = 1$. Say $u = w$. Notice that $x^3 + w \cdot y^3 = x^3 + y^3 + \lambda \cdot y^3$. The term on the left is divisible by λ^2, and $\lambda \cdot y^3$ is not divisible by λ^2. Therefore, $x^3 + y^3$ is not divisible by λ^2. But $x^3 + y^3 \equiv (x + y)^3 \bmod 3$, so that $x + y$ is divisible by λ, whence $(x + y)^3$ is divisible by λ^2, a contradiction. A similar argument shows that u can't be w^2.

Next we claim that $e \geq 2$. Switching x, y if necessary, we can write $x = a \cdot \lambda - 1$ and $y = b \cdot \lambda + 1$ and compute

$$x^3 + y^3 = (a^3 + b^3) \cdot \lambda^3 + 3(b^2 - a^2) \cdot \lambda^2 + 3(b + a) \cdot \lambda$$

Taking this mod λ^4 we get

$$-3 \cdot w \cdot \lambda \cdot (a^3 + b^3) + 3 \cdot \lambda \cdot (b + a) = 3 \cdot \lambda \cdot (b - w \cdot b^3 + a - w \cdot a^3)$$

We have $b \equiv w \cdot b^3 \bmod \lambda$, and so λ^4 divides $x^3 + y^3$.

Now write

$$-v \cdot \lambda^{3e} \cdot z^3 = x^3 + y^3 = (x + y)(x + w \cdot y)(x + w^2 \cdot y)$$

Since $e \geq 2$, λ^2 divides one of the three factors on the right. We can assume that λ^2 divides $x + y$. (For instance, if λ^2 divides $x + w \cdot y$, replace y by $w \cdot y$.)

We claim that λ divides $x + w \cdot y$ but that λ^2 does not divide this number. Indeed, $x + w \cdot y = x + y + \lambda \cdot y$ and this shows that λ divides. If λ^2 divides $x + w \cdot y$, then the same equation shows that λ^2 divides $\lambda \cdot y$, so that λ divides y. Since λ divides $x + y$, it follows that λ divides x, contradicting the fact that x, y have no common prime divisors.

We claim that λ is the only prime divisor of $x + y$ and $x + w \cdot y$. Indeed, if π divides $x + y$ and $x + w \cdot y$, then π divides $\lambda \cdot y$. If π is not an associate of λ, then π divides y. Already π divides $x + y$, and now we see that π divides x, a contradiction.

We can make the same two claims regarding $x + y$ and $x + w^2 \cdot y$ and regarding $x + w \cdot y$ and $x + w^2 \cdot y$.

We have that λ^{3e} divides the product of three factors. The numbers $x + w \cdot y$ and $x + w^2 \cdot y$ have exactly one factor of λ; thus, $x + y$ has a factor of λ^{3e-2}. Each prime divisor of k is a prime divisor of exactly one of $x + y$ and $x + w \cdot y$ and $x + w^2 \cdot y$. Thus, the primes in k^3 arrange themselves in cubes in each of the three numbers. Taking any necessary units into account, we see that there are $a, b, c \in \mathbb{Z}[w]$ and units p, q, r in $\mathbb{Z}[w]$ such that

$$x + y = \lambda^{3e-2} \cdot a^3 \cdot p$$
$$x + w \cdot y = \lambda \cdot b^3 \cdot q$$
$$x + w^2 \cdot y = \lambda \cdot c^3 \cdot r$$

Compute

$$(x + y) + w \cdot (x + w \cdot y) + w^2 \cdot (x + w^2 \cdot y) = 0$$
$$\lambda^{3e-2} \cdot a^3 \cdot p + \lambda \cdot b^3 \cdot q + \lambda \cdot c^3 \cdot r = 0$$

Dividing out by λ, we obtain an equation of the type of the counterexample but with a smaller value of e. This is a contradiction. $\square$

Notice that the equation $x^3 + y^3 = z^3$ can be written $x^3 + y^3 + (-z)^3 = 0$.

THEOREM 4.5. *Let x, y, z be non-zero integers. Then $x^3 + y^3 + z^3 \neq 0$.*

PROOF. Assume that $x^3 + y^3 + z^3 = 0$. We can assume that x, y, z have no common prime factor.

The integers x, y have no common prime divisor, since if the integer prime p divides both x and y, then it divides z, a contradiction. The GCD Theorem finds integers a, b such that $a \cdot x + b \cdot y = 1$. If follows that x, y have no common prime divisor in $\mathbb{Z}[w]$, for a common prime divisor would have to divide 1 (and be a unit!).

Next we claim that 3 has to divide one of x, y, z. Suppose that 3 divides none of them. Then $x^3 + y^3 + z^3 = 0$ holds for x, y, z in U_9. A case by case check in U_9 shows that this is not possible.

Say 3 divides z, and write $z = 3^k \cdot j$ where $k \geq 1$ and 3 does not divide j. In $\mathbb{Z}[w]$, we have $z = \lambda^{2k} \cdot j$ and λ does not divide j. Thus, $z^3 = \lambda^{6k} \cdot j^3$. The equation $x^3 + y^3 + \lambda^{6k} \cdot j^3 = 0$ violates Lemma 4.4. $\qquad \square$

4. Pythagorean Triples

The main theorem in this section can be proved by methods more naive than the theory we have been developing, but the subject fits very naturally with the other topics in this chapter, and so we include it here. A *Pythagorean triple* consists of positive integers a, b, c such that $a^2 + b^2 = c^2$. Given a Pythagorean triple a, b, c and a positive integer q, it is easy to see that qa, qb, qc is a Pythagorean triple. Thus, we are naturally interested in the triples which have no common prime divisor. We call such a triple a *coprime triple*. Another Diophantine problem: find all the coprime triples.

The following result goes back to Euclid.[4]

PROPOSITION 4.6. *Let m, n be positive integers not both odd and with GCD 1. Assume that $m > n$. Then $m^2 - n^2, 2mn, m^2 + n^2$ is a coprime triple. Every coprime triple has this form.*

PROOF. Given $m > n$, not both odd and with GCD 1, we leave it to you to show that the three numbers given form a coprime triple.

---

[4]See [**7**, Book X, Proposition 29].

Assume that a, b, c form a coprime triple. If a, b had a common prime factor, then that number would also divide c, a contradiction. Thus, a, b have GCD equal to 1, and so there are integers x, y with

(4.5) $x \cdot a + y \cdot b = 1$

We can factor $c^2 = (a+bi)(a-bi)$. Then $a+bi$ is not a unit in $\mathbb{Z}[i]$ (Why?), and so there is a prime π in $\mathbb{Z}[i]$ that divides $a + bi$. We will show that the exponent of π as a factor of $a + bi$ is even. If not, then since its exponent as a factor of c^2 must be even, we see that π divides both $a + bi$ and $a - bi$. Then also, π divides $2a$ and $2bi$, so that π divides $2b$. Multiplying equation (4.5) by 2, we see that π divides 2, and therefore π is an associate of $1 + i$.

Recall that π divides c. We conclude that 2 divides c, and so 4 divides $a^2 + b^2$. We have seen that this implies that a, b are both even. This is a contradiction.

We have established that if π divides $a + bi$, then it divides it an even number of times. It follows that $a + bi$ has the form $u \cdot (m + ni)^2$, for some unit u in $\mathbb{Z}[i]$. Multiplying out the square, and using that $u \in \{\pm 1, \pm i\}$, we see that a, b, c have the form claimed.

We claim that m, n have GCD 1. Indeed, if they had a common prime factor, it would divide a, b, c. Finally, m, n are not both odd, for if they were, we see that a, b, c would all be even. $\square$

5. Fermat's Last Theorem: Exponent 4

Our final Diophantine problem: there are no nonzero integers x, y, z such that $x^4 + y^4 = z^4$. The proof we present is more tricky than difficult. Undoubtedly, many people have discovered similar arguments. The author discovered it by playing with coprime triples one sunny day.

THEOREM 4.7. *There are no nonzero integers x, y, z such that $x^4 + y^4 = z^4$.*

PROOF. We will show that the equation $x^4 + y^4 = w^2$ has no nonzero integer solutions. The theorem will follow, for we can let $w = z^2$ in the equation of the theorem.

To get a contradiction, choose a solution of positive numbers x, y, w with w as small as possible. We claim that x, y have no common prime factor. Assume, to the contrary, that p is a common prime to x, y. Then p^4 divides $x^4 + y^4$, and so p^4 divides w^2. It follows that p^2 divides w. Now, the integers $x/p, y/p, w/p^2$ satisfy the equation in question but with a smaller w-term. This contradicts the minimality of w, and we conclude that x, y have no common prime factor.

We will apply Proposition 4.6 twice to produce a solution with a smaller w. Along the way we will obtain two inequalities (4.8) and (4.11) which will force a contradiction, but we advise that you ignore these inequalities on first reading.

We have $(x^2)^2 + (y^2)^2 = w^2$, and x^2, y^2, w are a coprime triple. We can assume that x^2 is odd and y^2 is even. Proposition 4.6 finds positive integers a, b such that

$$(4.6) \qquad\qquad x^2 = a^2 - b^2$$

$$(4.7) \qquad\qquad y^2 = 2ab$$

We also have $w = a^2 + b^2$, and so

$$(4.8) \qquad\qquad a < w$$

Rewrite (4.6) as $x^2 + b^2 = a^2$. Since a, b have no common prime factor, x cannot have a prime factor in common with either of them. Thus x, b, a are a coprime triple. We already know that x is odd, and Proposition 4.6 then forces that b is even, and it finds positive integers c, d such that

$$(4.9) \qquad\qquad a = c^2 + d^2$$

(4.10)
$$b = 2cd$$

Notice also that a is odd.

Now take the information that b is even and that a, b have no common prime factor back to equation (4.7). A prime factor of a is an odd prime factor of y, and therefore occurs in y^2 with even exponent (as a square). Since a, b have no common primes, this prime does not occur in b, and we conclude that the prime occurs in a with even exponent. It follows that a is a square: $a = e^2$, and we have

(4.11)
$$e \leq a$$

Substituting $a = e^2$ into (4.7) we see that $2b$ is an integer square. Since b is even, we can then write

(4.12)
$$b = 2f^2$$

for some integer f.

We substitute equation (4.12) into equation (4.10) to obtain that $f^2 = cd$. Because c, d are relatively prime, it follows that both c and d are squares:

(4.13)
$$c = g^2$$

(4.14)
$$d = h^2$$

Now substitute (4.13), (4.14), and $a = e^2$ into (4.9) to get

$$e^2 = g^4 + h^4$$

This is a solution to the original equation. The two inequalities (4.8) and (4.11) show that $e < w$, and this is a contradiction to the choice of w as minimal. □

The method of the previous proof replaces the solution to an equation by a smaller solution. This ancient technique is called the *Method of Descent*. There are many problems that can be solved by this method.

6. Problems

1. Let $I = \sqrt{2} \cdot \mathbb{Z}[\sqrt{2}]$. Show that $\mathbb{Z}[\sqrt{2}]/I$ is ring isomorphic to $\mathbb{Z}_2$.

2. For each of the quadratic PID's listed below and for each of the integer primes $2, 3, 5, 7$, figure out whether the prime ramifies, splits, or extends in each of the rings.

$$\mathbb{Z}[\sqrt{3}], \quad \mathbb{Z}[\sqrt{-2}], \quad \mathbb{Z}[(1 + \sqrt{5})/2], \quad \mathbb{Z}[(1 + \sqrt{-11})/2]$$

3. Show that the units of $\mathbb{Z}[\sqrt{3}]$ are the integer powers of $2 + \sqrt{3}$, times ± 1.

(a) Show that $2 + \sqrt{3}$ is a unit.

(b) Let $\alpha = a + b \cdot \sqrt{3}$ be a unit. Assume that $b = 0$ and get the conclusion.

(c) Assume that $a > 0$ and $b \neq 0$ and look at $\alpha \cdot (2 + \sqrt{3})$ and $\alpha \cdot (2 - \sqrt{3})$. Show that one of them has a smaller "a" term.

(d) Use induction to get the result for the case $a > 0$.

(e) Do the cases $a = 0$ and $a < 0$.

4. Let $n \geq 2$ be an integer. Show that the units of $\mathbb{Z}[\sqrt{-n}]$ are ± 1.

5. Complete the following, more elementary proof of Proposition 4.6. Let a, b, c be a coprime Pythagorean triple.

(a) Show that a, b cannot both be even, and they cannot both be odd.

(b) Assume that a is even and b odd. Write $a^2 = (c - b) \cdot (c + b)$. Show that $c - b$ and $c + b$ have GCD equal to 2.

(c) Show that there are integers m, n such that $c + b = 2 \cdot m^2$ and $c - b = 2 \cdot n^2$.

(d) Show that $c = m^2 + n^2$ and $b = m^2 - n^2$ and $a = 2 \cdot m \cdot n$.

6. Show that in $\mathbb{Z}[\sqrt{2}]$, there are a, b, c with $a^2 + b^2 = c^2$, for which there are no elements $m, n \in \mathbb{Z}[\sqrt{2}]$ such that $c = m^2 + n^2$. (Thus, Proposition 4.6 is false in the ring $\mathbb{Z}[\sqrt{2}]$.)

CHAPTER 5

Polynomial Rings

1. Coefficients, Degree, Factors

Let R be a commutative ring, and we write $R[X]$ for the set of *polynomials*[1] with coefficients in R (we often say, polynomials *over* R). A polynomial $f(X)$ over R can be written

$$f(X) = \sum_{k=0}^{n} f_k \cdot X^k \quad \text{where} \quad f_0, \ldots, f_n \in R$$

The element f_k is the *coefficient* of X^k. Notice that we are using the same letter for the coefficient sequence as for the polynomial itself. We denote polynomials by such notation as $f(X)$ to emphasize the possible presence of X, but we must remember that the elements of R are themselves polynomials – they are the constant polynomials. If g is a constant polynomial, then $g = g_0$.

You know that the n at the top of the summation is not unique, since, for instance

$$2 + X = 2 + X + 0X^2$$

We assume you know how to add and multiply polynomials. We recall the degree of a polynomial: If $f(X)$ is a non-zero polynomial, then some coefficient is non-zero. The *degree* of $f(X)$ is the maximal positive integer n such that $f_n \neq 0$. For instance, if $f(X)$ is a non-zero element of R, then its degree is

---

[1]We assume that you have seen a construction of the polynomial ring, or that you are willing to let your familiarity with real polynomials carry over. We can show you a formal construction, if you wish. The X is an *indeterminate* – a formal symbol, as introduced on p.33.

0. In all cases, we write $\deg(f(X))$ for the degree of f. The degree of the 0 polynomial is problematic, and so we leave it undefined. In a domain, the degree has a logarithmic property.

PROPOSITION 5.1. *Let R be a domain. Let $f(X)$ and $g(X)$ be non-zero elements of $R[X]$. Then $\deg(f(X)g(X)) = \deg(f(X)) + \deg(g(X))$.*

PROOF. Let $n = \deg(f(X))$ and write $f(X) = \sum_{k=0}^{n} f_k \cdot X^k$, where $f_n \neq 0$. Similarly, let $m = \deg(g(X))$, and write $g(X) = \sum_{j=0}^{m} g_j \cdot X^j$. Then

$$f(X) \cdot g(X) = \sum_{k=0}^{n} \sum_{j=0}^{m} f_k \cdot g_j \cdot X^{k+j}$$

We see that the highest power of X involved is X^{n+m}, and this term has coefficient $f_n \cdot g_m$. Since R is a domain and $f_n \neq 0 \neq g_m$, this coefficient is not zero.   □

You will have an exercise to show that Proposition 5.1 is false when R is not a domain.

COROLLARY 5.2. *If R is a domain, then $R[X]$ is a domain. The units of $R[X]$ are the units of R.*

PROOF. Proposition 5.1 shows that if $f(X), g(X)$ are non-zero, then $f(X) \cdot g(X)$ is non-zero. Thus, $R[X]$ is a domain.

Units? Let $f(X), g(X) \in R[X]$ with $f(X) \cdot g(X) = 1$, then of course neither is 0. Proposition 5.1 applies to show that $\deg(f(X)) + \deg(g(X)) = 0$, from which it follows that f and g must be elements of R.   □

In our later work on fields, we will need the following, which says that we can extend a ring homomorphism to the corresponding polynomial rings.

PROPOSITION 5.3. *Suppose that $\alpha : R \to S$ is a ring homomorphism, where R, S are domains. Then there is a unique ring homomorphism $\beta : R[X] \to S[X]$ such that $\beta(r) = \alpha(r)$ for all $r \in R$, and with $\beta(X) = X$.*

PROOF. Define

$$\beta \left(\sum_{k=0}^{n} a_k \cdot X^k \right) = \sum_{k=0}^{n} \alpha(a_k) \cdot X^k$$

Since $\alpha(0) = 0$, it doesn't matter if n is the degree or not. It is a direct calculation that β is a ring homomorphism, and we observe from its definition that it has the required properties. □

Next, we want to view polynomials as functions, "plugging in ring elements." The best way to understand this is to see that evaluation is a ring homomorphism.

Let R be a commutative ring, and let $f(X) \in R[X]$. Write

$$f(X) = \sum_{k=0}^{n} f_k \cdot X^k$$

For $a \in R$, we define

$$f(a) = \sum_{k=0}^{n} f_k \cdot a^k$$

We say that we *evaluate* $f(X)$ at $X = a$. The proof of the following is an exercise.

PROPOSITION 5.4. *Let R be a commutative ring, and let $a \in R$. The function $\sigma : R[X] \to R$ defined by $\sigma(f(X)) = f(a)$ is a ring homomorphism. The image $\sigma(R[X])$ is a subring of R.*

We have to be careful when we regard polynomials as functions. As a function on $\mathbb{Z}_2$, the polynomial $X^2 + X$ is *constant* – plug in $X = 0, 1$, and you get 0 both times. Yet $X^2 + X$ is not a constant polynomial. On the other hand, one can prove that if $f(X) \in \mathbb{Q}[X]$ and $f(a)$ is constant for all $a \in \mathbb{Q}$, then $f(X)$ is a constant polynomial. In analysis it is common to regard polynomials as functions; in algebra they are elements of a ring. Proposition 5.4 gives a ring-theoretic interpretation of polynomials as functions.

There is another analytic idea that transfers to algebra: we will find it convenient to have the derivative! For a commutative ring R, we define the *derivative* $f'(X)$ for $f(X) \in R[X]$ by the formula

$$\left(\sum_{k=0}^{n} a_k \cdot X^k \right)' = \sum_{k=1}^{n} a_k \cdot k \cdot X^{k-1}$$

Notice that we don't use any limits here – they wouldn't make sense in an arbitrary ring. Also, the sum in the derivative starts at $k = 1$, since the constant term a_0 is lost. This gives the familiar fact that the derivative of a constant is 0. The expression $k \cdot X^{k-1}$ in $f'(X)$ should really be $k * X^{k-1}$ since the *integer* k might not be an element of R. We will abuse notation and avoid the asterisk. The exercises show that the derivative has the properties with which you are familiar from calculus.

Now we wish to consider polynomials over a field. Recall Theorem 3.6, which says that a domain with a Euclidean norm is a PID.

PROPOSITION 5.5. *Let F be a field, let $f(X), g(X) \in F[X]$ and suppose that $g(X) \neq 0$. Then there are $q(X), r(X) \in F[x]$ with*

$$f(X) = q(X)g(X) + r(X)$$

and either $r(X) = 0$ or $\deg(r(X)) < \deg(g(X))$.

PROOF. Polynomial long division! We give a formal induction argument to show that it works for an arbitrary field F. Write $m = \deg(g(X))$, and we use induction on n.

If $f(X) = 0$ or if $\deg(f) < m$, then

$$f(X) = 0 \cdot g(X) + f(X)$$

satisfies our requirements.

Now assume that $n = \deg(f(X))$, that $n \geq m$. Write $f(X) = \sum_{k=0}^{n} f_k \cdot X^k$ and $g(X) = \sum_{j=0}^{m} g_j \cdot X^j$, and compute that

$$h(X) = f(X) - \frac{f_n}{g_m} \cdot X^{n-m} \cdot g(X)$$

has degree at most $n-1$. Using $h(X)$ in place of $f(X)$, induction on the degree shows that there are polynomials $q(X), r(X)$ with $r(x) = 0$ or $\deg(r(X)) < m$, and

$$h(X) = q(X) \cdot g(X) + r(X)$$

and it follows from the definition of $h(X)$ that

$$f(X) = \left(\frac{f_n}{g_m} \cdot X^{n-m} + q(X) \right) \cdot g(X) + r(X)$$

as needed. □

It follows that $F[X]$ has a Euclidean norm, with all the trimmings.

PROPOSITION 5.6. *Let F be a field and define $N : F[X] \to \mathbb{N}$ by $N(0) = 0$ and $N(f(X)) = 1 + \deg(f(X))$ for all $f(X) \neq 0$. Then N is a Euclidean norm. Thus, $F[X]$ is a PID and a UFD.*

PROOF. Proposition 5.5 completes the proof that N is a Euclidean norm. Theorem 3.6 and Theorem 3.9 then lead to PID and UFD. □

What are the primes (irreducibles) of $F[X]$? What are the associates? What are the units? Corollary 5.2 answers the last question: the units are the units of F, that is to say the nonzero elements of F. Therefore, two elements of $F[X]$ are then associates if and only if they are nonzero constant multiples of each other. A non-zero polynomial $f(X)$ of degree n is *monic* if its X^n

coefficient is 1. If $f(X)$ is nonzero of degree n, then $f(X)/f_n$ is the unique monic associate of $f(X)$. If we use the set of monic irreducibles as a designated set of primes, then Theorem 3.5 proves the following.

THEOREM 5.7. *Let $f(X) \in F[X]$ with degree $n > 0$. Let $g_1(X), \dots, g_m(X)$ be all the distinct monic irreducible divisors of $f(X)$. Then there are unique positive integers e_i such that $f(X) = f_n \cdot g_1(X)^{e_1} \dots g_m(X)^{e_m}$.*

Given $a \in F$, what does Proposition 5.5 say about division by $X - a$?

REMAINDER THEOREM. *Let $f(X) \in F[X]$ and let $a \in F$ where F is a field. Then $f(X) = q(X)(X - a) + f(a)$ where $q(X) \in F[X]$.*

PROOF. Proposition 5.5 finds $q(X), r(X) \in F[X]$ such that either $r(X) = 0$ or $\deg(r(X)) < \deg(X - a)$, and with $f(X) = q(X)(X - a) + r(X)$. The requirement $r(X)$ shows that it is constant. Proposition 5.4 shows that evaluation at a is a ring homomorphism. It follows that

$$f(a) = q(a)(a - a) + r(a) = r(a)$$

Since $r(a) = r(X)$, we are done.   □

What about the case where $f(a) = 0$? (The case that a is a *root* of $f(X)$.) The answer is a mere corollary of the Remainder Theorem, although it carries its own name.

FACTOR THEOREM. *Let $f(X) \in F[X]$ where F is a field. Let $a \in F$, then $X - a$ is a factor of $f(X)$ if and only if $f(a) = 0$.*

Another familiar fact.

COROLLARY 5.8. *Let F be a field and $f(X)$ be nonzero in $F[X]$ of degree n. Then $f(X)$ has no more than n distinct roots in F.*

PROOF. If $f(X)$ has m distinct roots in F, say $a_1, \ldots a_m$, then $f(X)$ has $X - a_1, \ldots, X - a_m$ as factors, by the Factor Theorem. The $X - a_i$ are non-associate irreducibles in $F[X]$, and so the Theorem 3.4 shows that $f(X)$ has $(X - a_1) \ldots (X - a_m)$ as a divisor. Proposition 5.1 then shows that $\deg(f(X))$ is at least m. This proves the corollary. $\qquad\square$

Here are two examples to warn you to be careful with this corollary. We need F in Corollary 5.8 to be a field: you should show that $X^2 - 1$ has four roots in the ring $\mathbb{Z}_8$.

Our second example is more subtle. Define Q to be the ring of elements of $\mathfrak{M}(2, \mathbb{Q})$ of the form

$$\begin{pmatrix} a & 0 \\ 0 & a \end{pmatrix} \quad \text{for} \quad a \in \mathbb{Q}$$

You should see that Q is isomorphic to $\mathbb{Q}$, and so Q is a field. It is easy to see that the polynomial

$$f(X) = X^2 - \begin{pmatrix} 1 & 0 \\ 0 & 1 \end{pmatrix} \quad \text{in} \quad Q[X]$$

has *infinitely many roots*! Indeed, you can compute that

$$\begin{pmatrix} a & 1 - a^2 \\ 1 & -a \end{pmatrix}^2 = \begin{pmatrix} 1 & 0 \\ 0 & 1 \end{pmatrix} \quad \text{for all} \quad a \in \mathbb{Q}$$

Why doesn't this contradict Corollary 5.8? Because the corollary requires the roots to exist *in the given field*. The matrix ring is not a field. The only roots listed above that are actually in Q are the expected ones (plus and minus the identity matrix).

Corollary 5.8 has an application to groups that occur in fields. Notice that the field in the hypothesis does not have to be finite.

THEOREM 5.9. *Let F be a field and let G be a finite subset of F that is a group under the multiplication of F. Then G is a cyclic group.*

PROOF. This proof is based on the argument Gauss used in [9] to prove that the $\mathbb{Z}_p$ are cyclic when p is a prime.

Temporary notation: for a group K and a positive integer m, let K_m be the set of elements of K of order m. If n is the order of K, then by Lagrange's Theorem, the set K is the disjoint union of the K_m with m dividing n. For each positive integer n, let $D[n]$ be the set of positive integers dividing n. Then

$$(5.1) \qquad n = |K| = \sum_{m \in D[n]} |K_m|$$

Let G be as in the hypothesis, with $|G| = n$, and let H be a cyclic group of order n (we can have $H = \mathbb{Z}_n$ if we wish). For each positive integer m dividing n we will show that $|G_m| \leq |H_m|$.

Indeed, given such an m, suppose first that G_m is empty. Then the inequality is trivial.

Now what if G_m is not empty? Let $y \in G_m$, so that y has order m, and the cyclic group $\langle y \rangle$ consisting of powers of y has order m. If $z \in \langle y \rangle$, then $z^m = 1$, so that z is a root of the polynomial $X^m - 1$. The set $\langle y \rangle$ then gives m roots of $X^m - 1$. Corollary 5.8 says that $X^m - 1$ cannot have more than m roots, and so if $z \in G$ and $z^m = 1$, then $z \in \langle y \rangle$ It follows that $G_m = \langle y \rangle_m$.

Since H is cyclic, we have $H = \langle h \rangle$ for some $h \in H$, and we know that h has order n. If m divides n, write $n = q \cdot m$, and we can compute that h^q has order m. The cyclic groups $\langle y \rangle$ and $\langle h^q \rangle$ of order m are isomorphic, and so they have the same number of elements of order m. This proves that $|G_m| = |H_m|$ in this case.

We have either than $|G_m| = 0$ or $|G_m| = |H_m|$, and so we have proved that $|G_m| \leq |H_m|$.

Now compute

$$n = |G| = \sum_{m \in D[n]} |G_m|$$

$$\leq \sum_{m \in D[n]} |H_m| = |H| = n$$

We see that the inequalities $|G_m| \leq |H_m|$ must be equalities. In particular, since $|H_n| > 0$, we have $|G_n| > 0$, and G has an element of order n. This implies that G is cyclic. $\qquad\square$

A corollary: the non-zero elements of a finite field form a cyclic group. For instance, the non-zero elements of $\mathbb{Z}_p$, where p is prime, form a cyclic group.

2. Polynomials over the Integers.

Since $\mathbb{Q}$ is a field, Proposition 5.6 says that $\mathbb{Q}[X]$ is a PID. What about the subring $\mathbb{Z}[X]$?

PROPOSITION 5.10. *Let $I = 2 \cdot \mathbb{Z}[X] + X \cdot \mathbb{Z}[X]$. Then I is a non-principal ideal of $\mathbb{Z}[X]$.*

PROOF. Suppose that $I = f(X) \cdot \mathbb{Z}[X]$. Since 2 is an element of I, we must have $2 = f(X)g(X)$ for some $g(X) \in \mathbb{Z}[X]$. By Proposition 5.1, $\deg(f(X)) + \deg(g(X)) = 0$, and it follows that f and g are constants. Without loss of generality f and g are positive constants, and then f is either 1 or 2.

If $f = 1$, then the definition of I shows that

$$f = 2 \cdot h(X) + X \cdot j(X) \quad \text{for some} \quad h(X), j(X) \in \mathbb{Z}[X]$$

Evaluating at $X = 0$, we see that

$$1 = f(0) = 2 \cdot h(0)$$

an impossible equation in the integers.

Thus $f = 2$, and so $X \in I$ shows that

$$X = f \cdot h(X) \quad \text{for some} \quad h(X) \in \mathbb{Z}[X]$$

Evaluating at $X = 1$, we obtain the impossibility $1 = 2 \cdot h(1)$. $\square$

Although $\mathbb{Z}[X]$ is a subring of $\mathbb{Q}[X]$, Proposition 5.10 warns us to be careful to distinguish between them. By the way, the ideal $2 \cdot \mathbb{Q}[X] + X \cdot \mathbb{Q}[X]$ must be principal; can you see how?

We see that $\mathbb{Z}[X]$ cannot be a PID, but we will show that it is a UFD anyway – an interesting fact in its own right, but it is also true that the facts we establish along the way will be useful later. For now we will show that integer primes are still prime in $\mathbb{Z}[X]$; observe that such integers are units in $\mathbb{Q}[X]$.

LEMMA 5.11. *Let p be an integer prime. Then p is a prime element of $\mathbb{Z}[X]$.*

PROOF. Let $f(X), g(X) \in \mathbb{Z}[X]$ with p dividing $f(X)g(X)$. Write

$$p \cdot h(X) = f(X)g(X)$$

where $h(X) \in \mathbb{Z}[X]$. We know that there is an onto ring homomorphism $r : \mathbb{Z} \to \mathbb{Z}_p$ (this is discussed in Chapter 3). Proposition 5.3 shows that r extends to a ring homomorphism R mapping $\mathbb{Z}[X]$ to $\mathbb{Z}_p[X]$. Observe for $j(X) \in \mathbb{Z}[X]$ that $R(j(X)) = 0$ if and only if p divides all the coefficients of $j(X)$, in other words if and only if p divides $j(X)$ as an element of $\mathbb{Z}[X]$.

The equation $p \cdot h(X) = f(X)g(X)$ tells us that

$$R(f(X)) \cdot R(g(X)) \equiv 0$$

We know that $\mathbb{Z}_p$ is a field, hence Corollary 5.2 makes $\mathbb{Z}_p[X]$ a domain. Thus either $R(f(X)) = 0$, in which case $f(X)$ is divisible by p, or $R(g(X)) = 0$, whence $g(X)$ is divisible by p. This proves the lemma. $\square$

We wanted to begin using more sophistication in our proofs, and so we used the ring homomorphism R in the previous. There is a more elementary proof that you might discover by letting j be maximal such that p does not divide f_j, and k maximal such that p does not divide g_k, and show that p does not divide $(f(X)g(X))_{j+k}$; that would contradict the fact that p divides $f(X)g(X)$.

Now we wish to identify when an integer polynomial is irreducible in $\mathbb{Q}[X]$. By definition, such a polynomial $f(X)$ is irreducible if it has no factorization $f(X) = g(X)h(X)$ where $g(X)$ and $h(X)$ are elements of $\mathbb{Q}[X]$ which are not zero and not units. Corollary 5.2 says that the units of $\mathbb{Q}[X]$ are the non-zero elements of $\mathbb{Q}$, and so *not being a unit* means *has degree at least 1*. The content of Gauss' Lemma is that such a polynomial is irreducible if it has no factorization $f(X) = g(X)h(X)$ where $g(X)$ and $h(X)$ are elements of $\mathbb{Z}[X]$ where $g(X)$ and $h(X)$ have degree at least 1. In practice it is easier to work in $\mathbb{Z}[X]$ than in $\mathbb{Q}[X]$, and so the practical use of Gauss' Lemma is that it makes it easier to establish that a given polynomial is irreducible. The following proof uses a clever trick; it is therefore rather hard to motivate.

GAUSS' LEMMA. *Let $f(X) \in \mathbb{Z}[X]$ and assume that $f(X) = g(X)h(X)$ where $g(X), h(X) \in \mathbb{Q}[X]$. Then there is a nonzero rational number r, such that $r \cdot g(X)$ and $h(X)/r$ are in $\mathbb{Z}[X]$. In particular, if $f(X)$ has no factorization $f(X) = g(X)h(X)$ with $g(X), h(X) \in \mathbb{Z}[X]$ of degree at least 1, then $f(X)$ is irreducible as an element of $\mathbb{Q}[X]$.*

PROOF. We can find positive integers m and n such that $m \cdot g(X)$ and $n \cdot h(X)$ are in $\mathbb{Z}[X]$, for example, let m be the product of all the denominators of the coefficients of $g(X)$ and let n be the analogous object for $h(X)$. Let S be the set of all positive integers s which can be written $s = rq$ for some $r, q \in \mathbb{Q}$ such that $r \cdot g(X)$ and $q \cdot h(X)$ are in $\mathbb{Z}[X]$. (Note that s is required to be an *integer*, whereas r, q can be fractions.) The product mn is an element of S, and so S is nonempty. By well-ordering S has a minimal element, call it s, and factor it $s = rq$ as in the definition of S. We claim that $s = 1$; this will force $q = 1/r$, and the proof will be complete.

If s is not 1, then because it is a positive integer, it has a prime divisor p. Then p divides $s \cdot f(X)$ which can be written

$$s \cdot f(X) = (r \cdot g(X)) \cdot (q \cdot h(X))$$

where the two factors in parenthesis are elements of $\mathbb{Z}[X]$. By Lemma 5.11, p divides one of these factors, say p divides $r \cdot g(X)$. This says that $r \cdot g(X)/p$ is an element of $\mathbb{Z}[X]$, hence if $t = s/p$, then $t = (r/p)q$ is a factorization that puts t in S. This contradicts the minimality of s, and we see that $s = 1$, as needed.   □

Now we will prove that $\mathbb{Z}[X]$ is a UFD by following a similar path as that to Theorem 3.9; we will first show that irreducibles are prime, and then that all nonzero, non-units factor into irreducibles. Showing that irreducibles are prime takes a couple of lemmas.

LEMMA 5.12. *Let $f(X) \in \mathbb{Z}[X]$ be a non-constant irreducible, and let $r \in \mathbb{Q}$ be such that $r \cdot f(X) \in \mathbb{Z}[X]$. Then r is an integer.*

PROOF. We may write $r = a/b$ where a,b are integers with 1 as their GCD. That $(a/b)f(X)$ has integer coefficients shows $a \cdot f_j/b$ is an integer for each j; in other words b divides each $a \cdot f_j$. Since a and b have GCD 1, we see that b divides each f_j. Therefore, if p is an integer prime divisor of b, then p divides each coefficient of $f(X)$, and therefore p divides $f(X)$. We have that p is prime in $\mathbb{Z}[X]$. That $f(X)$ is irreducible now forces that $f(X)$ and p are associates, which cannot be, since $f(X)$ is not constant.

This contradiction shows that r is an integer.   □

LEMMA 5.13. *Let $f(X) \in \mathbb{Z}[X]$ be an irreducible. Then $f(X)$ is a prime in $\mathbb{Z}[X]$.*

PROOF. If $f(X)$ is constant, then Lemma 5.11 takes care of the proof. Let $f(X)$ be non-constant, and suppose that it divides the product $g(X)h(X)$ with $g(X), h(X) \in \mathbb{Z}[X]$. Gauss' Lemma says that $f(X)$ is irreducible as an element of $\mathbb{Q}[X]$, and so it is prime as an element of $\mathbb{Q}[X]$ (this does not complete the proof!). Thus $f(X)$ divides $g(X)$ or it divides $h(X)$ in $\mathbb{Q}[X]$. Without loss of generality assume that $f(X)$ divides $g(X)$; find $j(X) \in \mathbb{Q}[X]$ such that

$$(5.2) \qquad g(X) = f(X)j(X)$$

Gauss' Lemma find a rational number r such that $r \cdot f(X)$ and $j(X)/r$ are elements of $\mathbb{Z}[X]$. By Lemma 5.12, r must be an integer, and so $j(X) = r \cdot (j(X)/r)$ is now seen to be in $\mathbb{Z}[X]$. Thus equation (5.2) is a factorization in $\mathbb{Z}[X]$, and therefore $f(X)$ divides $g(X)$. $\qquad \square$

THEOREM 5.14. *The ring $\mathbb{Z}[X]$ is a UFD.*

PROOF. Let $f(X)$ be an element of $\mathbb{Z}[X]$ which is not zero and not a unit. We must show that $f(X)$ factors into primes. If $f(X)$ is a constant, then it factors into integer primes, which, by Lemma 5.11, are primes in $\mathbb{Z}[X]$, and we are done.

If $f(X)$ has degree at least 1, and if $f(X)$ is irreducible as an element of $\mathbb{Q}[X]$, then by Gauss' Lemma it is irreducible as an element of $\mathbb{Z}[X]$. Lemma 5.13 says that $f(X)$ is prime, and we are again done.

If $f(X)$ is not irreducible in $\mathbb{Q}[X]$, then it factors $f(X) = g(X)h(X)$ where $g(X)$ and $h(X)$ are non-units in $\mathbb{Q}[X]$. In particular, $g(X)$ and $h(X)$ have degree at least 1. By Gauss' Lemma, there is a factorization $f(X) = a(X)b(X)$ where $a(X), b(X) \in \mathbb{Z}[X]$ have the same degrees as $g(X), h(X)$.

By induction on $\deg(f(X))$, $a(X)$ and $b(X)$ factor into primes, and we are completely done. $\square$

We mention that another route to Theorem 5.14 is to prove that $\mathbb{Z}[X]$ is noetherian, and then to use Lemma 5.13. The proof that $\mathbb{Z}[X]$ is noetherian is made difficult by the fact that $\mathbb{Z}[X]$ is not a PID.

3. Eisenstein's Criteria

There is an extremely easy way to create irreducible elements of $\mathbb{Q}[X]$ that have integer coefficients.

EISENSTEIN'S CRITERIA. *Let $f(X) \in \mathbb{Z}[X]$, have degree $n \geq 1$. Let p be an integer prime. Suppose that*

(a) *p does not divide f_n;*

(b) *p divides f_i, for $0 \leq i < n$;*

(c) *p^2 does not divide f_0.*

Then $f(X)$ is irreducible in $\mathbb{Q}[X]$.

PROOF. By Gauss' Lemma it suffices to show the impossibility of a factorization $f(X) = g(X)h(X)$ where $g(X)$ and $h(X)$ are non-constant elements of $\mathbb{Z}[X]$. Indeed, assume that such a factorization exists.

Now $f_0 = g_0 h_0$, and conditions (b) and (c) say that this is divisible by p but not by p^2. Thus exactly one of g_0 and h_0 is divisible by p; say that p divides g_0. We record the fact:

(5.3) p does not divide h_0

We claim that p cannot divide all the coefficients of $g(X)$. If it did, then the factorization of $f(X)$ establishes the implication from p divides $g(X)$ that p divides $f(X)$. This last statement is false, by condition (a).

We have shown that p cannot divide all the g_k. Let k be chosen maximal such that p does divide $g_0, \ldots, g_k$. The last two sentences show that

(5.4) p does not divide g_{k+1}

We now claim that $k < n - 1$. For if m is the degree of $g(X)$ and r is the degree of $h(X)$, then, recalling that $n = \deg(f(X))$ and using Proposition 5.1, we see that $r + m = n$. Since g and h have degree at least one, we have that $m < n$. Now also, $f_n = g_m h_r$, and by property (a) this is not divisible by p. Therefore g_m is not divisible by p, and thus $k < m$. The two integer inequalities $k < m < n$ establish that $k < n - 1$.

We now know that $k + 1 < n$, and so condition (b) shows that f_{k+1} is divisible by p. Observe that

$$f_{k+1} = \sum_{j=0}^{k+1} g_j h_{k+1-j}$$

For $j \le k$, we have that p divides g_j, and so p divides $g_j h_{k+1-j}$. The only other term in the summation is $g_{k+1} h_0$; in order for f_{k+1} to be divisible by p, we must have $g_{k+1} h_0$ divisible by p. But statements (5.3) and (5.4) show that $g_{k+1} h_0$ is not divisible by p. This contradiction completes the proof. $\square$

As a first use of Eisenstein's Criteria, we study $X^p - 1$ when p is a positive integer prime. We have

$$X^p - 1 = (X - 1) \cdot (1 + X + \cdots + X^{p-1})$$

The p-th *cyclotomic polynomial* $\Phi_p(X)$ is defined to be the rightmost factor:

$$\Phi_p(X) = 1 + X + \ldots + X^{p-1}$$

and we have

(5.5) $X^p - 1 = (X - 1) \cdot \Phi_p(X)$

This equations shows that the complex roots of $\Phi_p(X)$ are the p-th roots of 1, except for 1 itself. DeMoivre's Theorem then says that if $z = \exp(2\pi i / p)$, then these p-th roots are the numbers $z, z^2, \ldots, z^{p-1}$. These roots show us how to factor $\Phi_p(X)$. Since that polynomial has leading coefficient 1, we have

$$(5.6) \qquad \Phi_p(X) = (X - z)(X - z^2) \ldots (X - z^{p-1})$$

We will work toward proving that each $\Phi_p(X)$ is irreducible. We begin with a substitution that will allow us to use Eisenstein's Criteria. We want to use Proposition 5.4: map $f(X) \in \mathbb{Q}[X]$ to $f(X + 1)$, plugging in the element $X + 1$ of $\mathbb{Q}[X]$. The proposition shows that the mapping that sends $f(X)$ to $f(X + 1)$ is a ring homomorphism from $\mathbb{Q}[X]$ to $\mathbb{Q}[X]$. It is easy to see that $\deg(f(X + 1)) = \deg(f(X))$.

LEMMA 5.15. *Let $f(X) \in \mathbb{Q}[X]$ and suppose that $f(X+1)$ is an irreducible element of $\mathbb{Q}[X]$. Then so is $f(X)$.*

PROOF. If $f(X) = g(X)h(X)$, then because substituting $X + 1$ is a ring homomorphism, we get $f(X + 1) = g(X + 1)h(X + 1)$. Since $f(X + 1)$ is irreducible, one of $g(X + 1), h(X + 1)$ is constant. If $g(X + 1)$ is constant, for example, then so is $g(X)$. $\qquad \square$

THEOREM 5.16. *If p is a positive integer prime, then $\Phi_p(X)$ is an irreducible element of $\mathbb{Q}[X]$.*

PROOF. We compute $\Phi_p(X + 1)$. Because of (5.5), we have

$$(X + 1)^p - 1 = \Phi_p(X + 1) \cdot (X + 1 - 1)$$

The Binomial Theorem writes this equation:

$$\sum_{k=1}^{p} \binom{p}{k} \cdot X^k = \Phi_p(X + 1) \cdot X$$

(Notice that the $k = 0$ term on the left was cancelled by the -1.) We are working in the domain $\mathbb{Q}[X]$, and so we can cancel X on both sides.

$$\Phi_p(X+1) = \sum_{k=1}^{p} \binom{p}{k} \cdot X^{k-1}$$

Proposition 1.2 shows that this polynomial meets Eisenstein's Criteria: $\Phi_p(X+1)$ is irreducible. Lemma 5.15 says that $\Phi_p(X)$ is irreducible. $\qquad\square$

Theorem 5.16 leads naturally to the question, "Given the positive integer n, what are the irreducible factors of $X^n - 1$?" The answer to this question is elegant and will be given in Chapter 8, Section 1.

4. Problems

1. Find a commutative ring R and polynomials $f(X)$ and $g(X)$ of degree 3 and 5, respectively, and with $\deg(f(X) \cdot g(X)) = 2$.

2. Let R be a commutative ring. For all $f(X), g(X) \in R[X]$ and $a \in R$, prove the following.

(a) $(a \cdot f(X))' = a \cdot f'(X)$
(b) $(f(X) + g(X))' = f'(X) + g'(X)$
(c) $(f(X) \cdot g(X))' = f'(X) \cdot g(X) + f(X) \cdot g'(X)$

(Hint for (c): do the case $f(X) = X^k$ and $g(X) = X^j$ first; then use the previous properties to establish the general case.)

3. Let F be a field, let $\alpha \in F$, and let $f(X) \in F[X]$. Show that $(X - \alpha)^2$ divides $f(X)$ if and only if $f(\alpha) = 0$ and $f'(\alpha) = 0$.

4. Suppose that F is a field of characteristic 0, and let $f(X) \in F[X]$ be irreducible. Show that $f'(X) \neq 0$.

5. Suppose that $f(X) \in \mathbb{Z}_p[X]$, where p is a positive integer prime, and suppose that $f'(X) = 0$. Show that there is $g(X) \in \mathbb{Z}_p[X]$ such that $f(X) = g(X^p)$.

6. Let F be a field. Show that $F[X]$ has infinitely many irreducibles. (Hint: mimic the proof that the integers has infinitely many primes.)

7. For a field F, the *power series ring* $P(F)$ is defined as the set of expressions

$$\sum_{k=0}^{\infty} f_k \cdot X^k \quad \text{where} \quad f_k \in F, \quad \text{for all} \quad k$$

This infinite sum is formal – there is no limit or calculus involved. Show how to define addition and multiplication on $P(F)$ by giving explicit formulas for the coefficients of a sum/product in each case. (It turns out that $P(F)$ is a ring under these operations; you do not have to prove that.)

8. Let F be a field, and assume that $P(F)$ is a ring. Suppose that $f \in P(F)$ has $f_0 \neq 0$. Show that f is a unit in $P(F)$. (Hint: solve for the coefficients of g such that $f \cdot g = 1$.)

9. Let F be a field, and assume that $P(F)$ is a ring. Define $X \in P(F)$ by $X_1 = 1$ and $X_k = 0$ for all $k \neq 1$. Show that X is prime in $P(F)$ and that every prime in the ring is an associate of X.

10. Factor $-2X^4 - 2X^3 + 4X^2 + 6X + 6$ into irreducibles over $\mathbb{Q}[X]$, over $\mathbb{R}[X]$, and over $\mathbb{C}[X]$. (Prove that the irreducibles are irreducible in each case.)

11. Consider the field $\mathbb{Z}_3[\alpha]$, where $\alpha^2 = 2$. Since this field is finite, Theorem 5.9 says that its non-zero elements form a cyclic group. Find a generator of that group.

12. Let $f(X) \in \mathbb{Z}[X]$ be monic, and let $g(X) \in \mathbb{Q}[X]$ be a monic, irreducible factor of $f(X)$. Show that $g(X) \in \mathbb{Z}[X]$. (Hint: Gauss' Lemma.)

13. In each case find $f(X) \in \mathbb{Z}[X]$ of degree 3, irreducible over the rationals. (Hint for both: think about the graph $y = f(x)$.)

(a) $f(X)$ has 3 real roots.

(b) $f(X)$ has exactly 1 real root.

14. A polynomial with integer coefficients is called *primitive* if there is no prime that divides all of its coefficients. Show that the product of two primitive polynomials is primitive.

15. Let $f(X) \in \mathbb{Z}[X]$ be monic, and suppose that $b \in \mathbb{Q}$ with $f(b) = 0$. Show that $b \in \mathbb{Z}$. (Hint: write b in lowest terms and clear denominators in the equation $f(b) = 0$.)

16. Use the trick of Theorem 5.16 to prove that $X^4 + 1$ and $X^6 + X^3 + 1$ are irreducible in $\mathbb{Q}[X]$.

17. Factor $X^6 - 1$ and $X^8 - 1$ and $X^9 - 1$ into irreducible factors in $\mathbb{Q}[X]$. (Hint: the previous problem is relevant.)

18. Show that $\mathbb{Z}[X]/(2\mathbb{Z}[X])$ is a domain but not a field. (Hint: think ideals.)

19. Let $f(X)$ be an irreducible element of $\mathbb{Q}[X]$. Show that there is an integer n such that $n \cdot f(X)$ is prime in $\mathbb{Z}[X]$.

20. Let $A(X), B(X) \in \mathbb{Z}[X]$, and suppose there is no prime in $\mathbb{Z}[X]$ that divides both of them. Show that the GCD of $A(X), B(X)$, in $\mathbb{Q}[X]$ is 1. (Hint: let $C(X)$ be an irreducible in $\mathbb{Q}[X]$ that divides both $A(X)$ and $B(X)$. Show that we can take $C(X) \in \mathbb{Z}[X]$ to get a contradiction.)

21. Let $A(X), B(X) \in \mathbb{Z}[X]$, and suppose there is no prime in $\mathbb{Z}[X]$ that divides both of them. Let P be the set of integer primes p that divide every number $A(n), B(n)$, for all $n \in \mathbb{Z}$. Show that P is finite. (Hint: clear denominators in the equation that gives the GCD of $A(X), B(X)$ in $\mathbb{Q}[X]$.)

22. Let p be a positive integer prime. Define

$$A(X) = \prod_{j=1}^{p}(X - j)$$

Let q be a positive integer prime with $q \leq p$. Show that q divides $A(n)$ for all $n \in \mathbb{Z}$. (Notice that there is no integer prime that divides $A(X)$ in $\mathbb{Z}[X]$.)

23. (Fermat's Last Theorem for polynomial rings.) Let F be a field of characteristic 0 and let $n \geq 3$ be an integer. Suppose there are non-zero $A, B, C \in F[X]$ such that

(5.7) $A^n + B^n = C^n$

Follow the steps to prove that there is $D \in F[X]$ and $a, b, c \in F$ such that

$$A = a \cdot D, \quad B = b \cdot D, \quad C = c \cdot D, \quad a^n + b^n = c^n$$

(a) Use induction on the largest degree of A, B, C. Show that, without loss of generality, we can assume that A, B, C have no prime factor in common.

(b) Suppose that $\deg(A) \geq \deg(B)$. Then $\deg(A) \geq \deg(C)$, as well. Let $d = \deg(A)$.

(c) Take the derivative of both sides of (5.7), and work to get an equation of the form $A^n \cdot G = B^n \cdot H$, for some $G, H \in F[X]$.

(d) Show that since A, B have GCD 1 (why?), we have that A^n divides H.

(e) Estimate the degree of H and get a contradiction.

24. Let p be a positive integer prime. Find $A, B, C \in \mathbb{Z}_p[X]$, each non-constant, with no common prime factor, and such that $A^p + B^p = C^p$. (Hint: Proposition 2.21.)

25. (Another construction of $\mathbb{C}$.) Show that $\mathbb{R}[X]/(X^2 + 1)\mathbb{R}[X]$ is ring isomorphic to $\mathbb{C}$. (Hint: if $J = (X^2 + 1)\mathbb{R}[X]$, then $X + J$ looks like i.)

CHAPTER 6

Field Extensions and Degree

As rings, fields are extremely simple: no primes, no non-trivial ideals. The interesting subject here is the relationship between two fields, one of which contains the other. It might appear that the phrase "one of which contains the other" already gives a sufficient description of the relationship, but what we are interested in is the way in which the larger field may be constructed from the smaller field by solving polynomial equations.

1. Linear Algebra, Degree

Before we can deal with polynomial equations, however, we need to obtain a measure of the "distance" from the smaller field to the larger one. We will need the idea of dimension from linear algebra. Here is a basic fact that you have probably seen before: a consistent system of linear equations with more variables than equations has at least two solutions. In the following proposition we will have m equations in n unknowns with $n > m$. Notice that we are working over an *arbitrary field*.

PROPOSITION 6.1. *Let F be a field, and let m and n be positive integers with $m < n$. Let $A[i,j] \in F$ for $1 \leq i \leq n$ and $1 \leq j \leq m$. Then there are elements $x_i \in F$ for $1 \leq i \leq n$, not all zero, such that*

$$\sum_{i=1}^{n} x_i \cdot A[i,j] = 0 \quad \text{for all} \quad m \quad \text{with} \quad 1 \leq j \leq m$$

113

PROOF. We will use what amounts to Gaussian elimination! Our proof is by induction on m. Assume first that $A[1, j] = 0$ for all j. (In other words the equations never use x_1.) Then let $x_1 = 1$ and put all the other x_i equal to 0, and we have the required solution.

Now assume that some $A[1, j] \neq 0$. (In other words x_1 actually occurs in equation number j.) By renumbering the equations, if necessary, we can assume that $j = 1$, so that $A[1, 1] \neq 0$. We can solve the first equation for x_1:

$$x_1 = -\sum_{i=2}^{n} \frac{A[i, 1]}{A[1, 1]} \cdot x_i$$

If $m = 1$, then this is the only equation. Since $n > m = 1$, we see that x_1 is not the only unknown. Choose arbitrary nonzero values for the other unknowns, and then get x_1 from the equation just written. This completes the case $m = 1$.

Let $m > 1$, and then we can use the equation just computed to substitute for x_1 in the equations with $j \geq 2$, and we obtain $m - 1$ equations in $n - 1$ unknowns (x_k with $k \geq 2$). Since $m - 1 < n - 1$, induction gives us a solution to the $m - 1$ equations with not all the unknowns equal to zero. Determining x_1 from the equation substituted gives us a solution to *all* the equations with not all the unknowns equal to zero. $\square$

Let $F \subseteq E$ be fields, where F is a subring of E. Then we say that F is a *subfield* of E, and we say that E is an *extension* of F. Throughout, because we will always have at least two fields on the table at any given time, you must keep it straight which elements come from which field. Let S be a finite subset of E; say it has n elements $e_1, \ldots, e_n$. We say that S *spans E over F* if every element x of E can be written

$$x = \sum_{i=1}^{n} x_i e_i \quad \text{for some} \quad x_i \in F$$

In the language of linear algebra, we are saying that x is a linear combination of the e_i. But the *linear combination* gets its scalars from F, so it is a linear combination *over* F.

We say that S is *linearly independent in E over F* if whenever we have $x_i \in F$ for $1 \le i \le n$ and

$$\sum_{i=1}^{n} x_i e_i = 0$$

then it must be the case that $x_i = 0$ for all i. This is linear independence from linear algebra. As with the span, the scalars must come from F. For instance $1, \sqrt{2}$ are linearly independent over $\mathbb{Q}$, but $1, \sqrt{2}$ are *not independent* over $\mathbb{R}$. Be sure you see why these last two statements are true.

A mnemonic: *spanning sets are large* (large enough to generate all elements of E), and *linearly independent sets are small* (small enough to have only one linear combination equal to 0). The following, which enforces this heuristic, captures the heart of the matter.

PROPOSITION 6.2. *Let E be an extension of the field F. Let $e_1, \ldots, e_m$ span E over F, and let $d_1, \ldots, d_n$ be linearly independent in E over F. Then $n \le m$.*

PROOF. Suppose that $n > m$, and we will derive a contradiction. Because the e_j span E, each d_k can be written as a linear combination of the e_j:

$$d_k = \sum_{j=1}^{m} A[k, j] \cdot e_j \quad \text{for some} \quad A[k, j] \in F$$

Because $n > m$, Proposition 6.1 finds $x_1, \ldots, x_n$ in F, not all 0, such that

$$\sum_{k=1}^{n} x_k \cdot A[k, j] = 0 \quad \text{for} \quad 1 \le j \le m$$

Compute

$$\sum_{k=1}^{n} x_k \cdot d_k = \sum_{k=1}^{n} x_k \cdot \left(\sum_{j=1}^{m} A[k,j] \cdot e_j \right) = \sum_{k=1}^{n} \sum_{j=1}^{m} x_k \cdot A[k,j] \cdot e_j$$

$$= \sum_{j=1}^{m} \left(\sum_{k=1}^{n} x_k \cdot A[k,j] \right) \cdot e_j = \sum_{j=1}^{m} 0 \cdot e_j = 0$$

The fact that x_k are not all 0 contradicts the fact that the d_k are independent.

□

Proposition 6.2 is very powerful. We can combine the two concepts of span and linear independence: if S is a subset of the field E such that S is independent over the subfield F and such that S spans E over F, then we say S is a *basis* for E over F. This is the idea of basis from linear algebra. We are thinking of F as the *base field* and E as a vector space over F. In our heuristic, a basis is both small and large, therefore it is *just right*. Intuitively there can only be one possible number of *just right* size for E and F. In other words, each basis for E over F should have the same number of elements. We will now prove this, but we pause to say that we are **not** proving that there is only one basis – we'll see examples shortly.

PROPOSITION 6.3. *Let E be an extension field of F, and suppose that E is spanned by a finite set over F.*

(a) *If S spans E over F, then S contains a subset that is a basis for E over F.*

(b) *If S is linearly independent in E over F, then S is contained in a basis for E over F.*

(c) *Every basis for E over F has the same number of elements.*

PROOF. For (a), let S span E over F. There is a subset T of S that spans E and that has a minimal number of elements. We claim that T is linearly independent, so that T is a basis. Number the elements of T: t_j for $1 \leq j \leq n$. If T is not linearly independent, there are $x_j \in F$, not all 0, such that

$$\sum_{j=1}^{n} x_j \cdot t_j = 0$$

We can renumber the t_j so that x_1 is not 0, and then

(6.1)
$$t_1 = -\frac{1}{x_1} \cdot \sum_{j=2}^{n} x_j \cdot t_j$$

It is easy to see that $t_2, \ldots, t_n$ spans E, since, in a linear combination of $t_1, \ldots, t_n$, we can use 6.1 to eliminate t_1 and obtain a combination of $t_2, \ldots, t_n$. The fact that E is spanned by $t_2, \ldots, t_n$ contradicts the minimality of T. Thus, T is a basis.

To prove (b), let S be linearly independent in E over F. By hypothesis, there is a finite set T that spans E over F. Proposition 6.2 shows that $|S| \leq |T|$. Thus, there is a maximal linearly independent set U that contains S. We claim that U spans E, so that U is a basis. If U does not span E, there is $e \in E$ that is not a linear combination of elements of U. We leave it to you to show that $U \cup \{e\}$ is linearly independent, and this is a contradiction.

For (c), if S and T are bases, then S spans E and T is linearly independent, so that Proposition 6.2 shows that $|S| \geq |T|$. On the other hand, S is linearly independent and T spans, and so $|S| \leq |T|$. Thus, $|S| = |T|$. $\square$

When E is an extension of the field F such that there is a basis for E over F, we say that E is a *finite extension* of F. We will write $|E : F|$ for the common size of all the bases. We refer to $|E : F|$ as the *degree* of E over F. We remark that our definition of basis necessarily implies that a basis is a finite set. It is possible to give the definition so as to allow a basis to have infinitely many elements; we will not need to do this, but you might give some thought to how one would define an infinite basis.

We have seen fields of the form $\mathbb{Q}[\alpha]$ where α is a root of a quadratic polynomial. In these examples, $|\mathbb{Q}[\alpha] : \mathbb{Q}| = 2$. For instance, the elements of $\mathbb{Q}[\sqrt{2}]$ have the form $a + b \cdot \sqrt{2}$ for $a, b \in \mathbb{Q}$. Thus, $1, \sqrt{2}$ spans $\mathbb{Q}[\sqrt{2}]$ over $\mathbb{Q}$. On the other hand, if $a + b \cdot \sqrt{2} = 0$, then $a = b = 0$, since $\sqrt{2} \notin \mathbb{Q}$. Thus, $1, \sqrt{2}$ is independent, and so it is a basis.

Let $c \in \mathbb{Q}$ with $c \neq 0$, then $c, \sqrt{2}$ is also a basis for $\mathbb{Q}[\sqrt{2}]$, as is $1, c\sqrt{2}$. There are infinitely many bases for this field.

How about an extension field with no basis: $\mathbb{Q} \subset \mathbb{R}$. We will use a set-theoretic sledge hammer! If $e_1, \ldots, e_n$ were a basis for $\mathbb{R}$ over $\mathbb{Q}$, then $\mathbb{R}$ would be countable,[1] since the set of all

$$\sum_{j=1}^{n} a_j \cdot e_j \quad \text{where} \quad a_1, \ldots, a_n \in \mathbb{Q}$$

is a countable set. But $\mathbb{R}$ is not countable, so $\mathbb{R}$ cannot be the span.

Believe it or not we want to consider the situation of three fields.

THEOREM 6.4. *Let E be an extension of the field F, and let D be an extension of the field E. Then D is a finite extension of F if and only if D is a finite extension of E and E is a finite extension of F. If D is a finite extension of F then $|D : F| = |D : E| \cdot |E : F|$.*

PROOF. This argument takes some time, but it really isn't hard. Let D have a basis over F. We will show that E has a basis over F. Indeed, because D is spanned by $|D : F|$ elements, Proposition 6.2 shows that there cannot be more than $|D : F|$ linearly independent elements of E over F. Let S be a subset of E that is linearly independent over F, and with S having as many elements as possible. If S does not span E over F, then there is $e \in E$ that is not a linear combination of elements of S. It follows that $S \cup \{e\}$ is linearly

---

[1]Countability is typically discussed somewhat briefly in a course in analysis. For the details, see [**10**] or Appendix A of [**12**].

independent over F, a contradiction.[2] We conclude that S is a basis for E over F.

Still assuming that D has a basis over F, we show that D has a basis over E. A spanning set for D over F is also a spanning set of D over E. Proposition 6.3 finds a basis as a minimal spanning set.

Now assume that D has a basis over E: $d_1, \ldots, d_n$, and let E have a basis over F: $e_1, \ldots, e_m$. We claim that the set $d_j \cdot e_k$ for all j, k is a basis for D over F. This will prove the remaining claims of this theorem.

Let $x \in D$, we find $x_1, \ldots, x_n \in E$ such that

$$x = \sum_{j=1}^{n} x_j d_j$$

Furthermore, if $x = 0$, then all the x_i have to be 0. In general, each x_j is a combination of the e_k: there are $A[j, k] \in F$ such that

$$x_j = \sum_{k=1}^{m} A[j, k] \cdot e_k$$

If $x_j = 0$, then $A[j, k] = 0$ for all k. Putting the summations together, we have

$$x = \sum_{j=1}^{n} \sum_{k=1}^{m} A[j, k] \cdot e_k \cdot d_j$$

and we see that the $e_k d_j$ span D over F. If $x = 0$, then as we have remarked, all x_j are 0, and so all the $A[j, k]$ are 0. This proves that the $e_k d_j$ are independent.

$\square$

2. Simple Extensions

We describe a particular way that field extensions come about, and we will prove that all finite extensions of fields which contain the rationals are of this sort.

---

[2]You might notice that this argument has been used twice. We have chosen not to isolate it as a proposition.

Our starting point will be an extension E of a field F. Let α be an element of E such that $f(\alpha) = 0$ for some nonzero $f(X) \in F[X]$. We say that α is *algebraic over* F. For example, if E is a finite extension of F, then every element of E is algebraic over F. To see this, let $\alpha \in E$, and notice that by Proposition 6.2, the list

$$1, \alpha, \alpha^2, \ldots$$

cannot all be linearly independent over F (at most $|E : F|$ of them can be independent). Let n be the smallest positive integer such that the list $1, \alpha, \ldots, \alpha^n$ is not independent over F; indeed, write

$$0 = f_0 \cdot 1 + f_1 \cdot \alpha + \ldots + f_n \cdot \alpha^n$$

where f_j are elements of F which are not all zero.

If we let $f(X) = f_0 + \ldots + f_n X^n$, then we see that $f(X)$ is a nonzero element of $F[X]$ such that $f(\alpha) = 0$; we mean that α is a root of $f(X)$. Thus α is algebraic over F.

For an algebraic α, we need a description of the set of elements of $F[X]$ which have α as a root.

PROPOSITION 6.5. *Let E be an extension of the field F, and let $\alpha \in E$ be algebraic over F. Then*

(a) there is a unique monic irreducible $f(X) \in F[X]$ such that $f(\alpha) = 0$,
(b) if $g(X) \in F[X]$ and $g(\alpha) = 0$, then $f(X)$ divides $g(X)$.

PROOF. Let I be the set of those elements of $F[X]$ having α as a root. It is an exercise to show that I is an ideal of $F[X]$. (Show that it is a kernel of a ring homomorphism.) The fact that α is a root of some non-zero polynomial shows that I is not zero. Since $F[X]$ is a PID, the ideal I can be written $f(X)F[X]$ for some $f(X) \in F[X]$. We may suppose $f(X)$ to be monic. Note further that $f(X)$ is not constant, since nonzero constants have no roots.

We see that $\deg(f(X)) > 0$, and we claim that $f(X)$ is irreducible. If $f(x) = g(x)h(x)$, then $0 = f(\alpha) = g(\alpha) \cdot h(\alpha)$. Because E is a domain, $g(\alpha) = 0$ or $h(\alpha) = 0$. Suppose that $g(\alpha) = 0$, then $g(X) \in I$ implies that $f(x)$ divides $g(x)$. Since $g(x)$ already divides $f(x)$, we see that f and g are associates. This suffices to prove $f(X)$ irreducible.

Let $g(X)$ be a monic irreducible element of $F[X]$ with α as a root, and observe that $g \in I$ so that f divides g. This proves that g and f are associates, and then they are equal since they are both monic. This proves (a), and then (b) also holds.   □

The $f(X)$ granted by Proposition 6.5 is called the *minimal polynomial of α over F*. The *degree of α over F* is the degree of its minimal polynomial. You have to know the field to get the minimal polynomial: the minimal polynomial of $\sqrt{2}$ over $\mathbb{Q}$ is $X^2 - 2$; the minimal polynomial for $\sqrt{2}$ over $\mathbb{R}$ is $X - \sqrt{2}$.

Again assume that E is an extension field of F. Let $\alpha \in E$ be algebraic over F. Recall from Proposition 5.4 that evaluation at $X = \alpha$ is a homomorphism from $F[X]$ to E, and that the image of this homomorphism is a subring of E. We call this subring $F[\alpha]$, and then by definition this subring is the set of $g(\alpha)$ for all $g(X) \in F[X]$. We will see that the structure of this subring is, in fact, quite simple.

We say that $F[\alpha]$ is obtained from F by *adjoining α to F*. We also say that $F[\alpha]$ is a *simple extension* of F. Theorem 6.6 gives the structure of this ring. The statement of this theorem contains a great deal of redundancy for the purpose of making it very clear what $F[\alpha]$ looks like. The proof is a tour of the theory of rings and polynomials.

THEOREM 6.6. *Let α be an algebraic element of the extension E of the field F. Let n be the degree of α over F. Then $F[\alpha]$ is a field with $\alpha \in F$ and $F \subseteq F[\alpha] \subseteq E$, and $|F[\alpha] : F| = n$. The field $F[\alpha]$ is the set of $g(\alpha)$ where $g(X) \in F[X]$ has degree less than n. The set $1, \alpha, \dots, \alpha^{n-1}$ is a basis of $F[\alpha]$.*

PROOF. First, note that $F \subseteq F[\alpha]$ since if $r \in F$ is considered as a constant polynomial, then $r(\alpha) = r$. Second, α is an element of $F[\alpha]$, since α is what you get when you evaluate the polynomial X at α.

Let $f(X)$ be the minimal polynomial of α over F. Then

$$\deg(f(X)) = n > 0$$

We already know that $F[\alpha]$ is a subring of E. To show that it is a field, the definition says we need to show that nonzero elements have multiplicative inverses. They certainly have inverses in the field E where all this is going on, but we need to show that $F[\alpha]$ contains its own inverses. Indeed, let $r(\alpha)$ be a nonzero element of $F[\alpha]$ where $r(X) \in F[X]$. Since $r(\alpha)$ is not zero and $f(\alpha) = 0$, we see that $r(X)$ is not divisible by $f(X)$. By Proposition 5.6, we know that $F[X]$ is a PID, and so Theorem 3.7, the irreducible $f(X)$ is prime in $F[X]$. Since the prime $f(X)$ does not divide $r(X)$, Theorem 3.10 finds $g(X), h(X) \in F[X]$ such that $g(X) \cdot r(X) + h(X) \cdot f(X) = 1$. Evaluating at α, we see that $g(\alpha) \cdot r(\alpha) = 1$. Thus, $g(\alpha) \in F[\alpha]$ is the inverse of $r(\alpha)$. This proves that $F[\alpha]$ is a field.

Now let $\beta \in F[\alpha]$, so that there is $g(X) \in F[X]$, for which $\beta = g(\alpha)$, by the definition of $F[\alpha]$. By Proposition 5.5 we can write $g(X) = f(X)q(X) + r(X)$ where $q, r \in F[X]$ and $\deg(r(X)) < n$. Evaluating at $X = \alpha$, we see that $g(\alpha) = r(\alpha)$. This proves that $F[\alpha]$ is the set of $r(\alpha)$ where $r(X)$ has degree less than n. This also shows that $F[\alpha]$ is spanned by $1, \alpha, \ldots, \alpha^{n-1}$, for the coefficients of the polynomial $r(X)$ are the coefficients of these powers of α in $r(\alpha)$.

The set $1, \alpha, \ldots, \alpha^{n-1}$ is also linearly independent, since if

$$\sum_{j=0}^{n-1} r_j \cdot \alpha^j = 0$$

then the polynomial

$$r(X) = \sum_{j=0}^{n-1} r_j \cdot X^j$$

has α as a root. Since $f(X)$ is the minimal polynomial for α, we see that $r(X)$ has to be the 0 polynomial, and so all the r_j are 0. We see that $|F[\alpha] : F| = n$. $\qquad\square$

Fields obtained by adjoining roots of the same irreducible polynomial are isomorphic. This fact will be crucial to the Galois theory in later chapters.

THEOREM 6.7. *Let E and L be extensions of F, let $f(X)$ be an irreducible element of $F[X]$, and let α and β be roots of $f(X)$ in E and L, respectively. Then there is a unique ring isomorphism $\sigma : F[\alpha] \to F[\beta]$ such that $\sigma(\alpha) = \beta$ and for which $\sigma(r) = r$ for all $r \in F$.*

PROOF. Let n be the degree of $f(X)$. Theorem 6.6 shows that each element of $F[\alpha]$ can be written $r(\alpha)$ for a unique polynomial $r(X) \in F[X]$ of degree less than n. Similarly, each element of $F[\beta]$ can be written uniquely $r(\beta)$. We define $\sigma : F[\alpha] \to F[\beta]$ by $\sigma(r(\alpha)) = r(\beta)$.

It is trivial to see that $\sigma(b + c) = \sigma(b) + \sigma(c)$ for all $b, c \in F[\alpha]$, and to see that $\sigma(b) = b$ for all $b \in F$, and to see that $\sigma(\alpha) = \beta$.

Let $b, c \in F[\alpha]$. Then there are polynomials $r(X)$ and $s(X)$ of degree less than n such that $b = r(\alpha)$ and $c = s(\alpha)$. Proposition 5.5 writes

$$r(X) \cdot s(X) = t(X) + q(X) \cdot f(X)$$

where $t(X), q(X) \in F[X]$ and the degree of $t(X)$ is less than n. Since α, β are roots of $f(X)$, we see that

$$r(\alpha) \cdot s(\alpha) = t(\alpha) \quad \text{and} \quad r(\beta) \cdot s(\beta) = t(\beta)$$

In light of this, we compute

$$\sigma(b \cdot c) = \sigma(r(\alpha) \cdot s(\alpha)) = \sigma(t(\alpha))$$

$$= t(\beta) = r(\beta) \cdot s(\beta) = \sigma(r(\alpha)) \cdot \sigma(r(\beta)) = \sigma(b) \cdot \sigma(c)$$

We have proved that σ is a ring homomorphism.

To see that σ is unique, suppose that δ is such a ring homomorphism. Let $r(X) \in F[X]$, and use the properties of δ to compute that $\delta(r(\alpha)) = r(\beta) = \sigma(r(\alpha))$. $\qquad\square$

When we adjoin a sequence of algebraic elements, the order of adjoining is immaterial.

PROPOSITION 6.8. *Let $K \subseteq L$ be fields. Let $\alpha \in L$ be algebraic over K, and let $\beta \in L$ be algebraic over $K[\alpha]$. Then β is algebraic over K, and $K[\alpha][\beta] = K[\beta][\alpha]$; this field is a finite extension of K.*

PROOF. Theorem 6.4 shows that $K[\alpha][\beta]$ is a finite extension of K, and so β is algebraic over K. Theorem 6.6 shows that $\beta \in K[\alpha][\beta]$, and so we have $K[\beta] \subseteq K[\alpha][\beta]$. Now $\alpha \in K[\alpha][\beta]$ and it is algebraic over $K[\beta]$, so now $K[\beta][\alpha] \subseteq K[\alpha][\beta]$. Similarly, we have $K[\alpha][\beta] \subseteq K[\beta][\alpha]$, and now we see that these fields are equal. $\qquad\square$

We will write $K[\alpha, \beta]$ for either of $K[\alpha][\beta]$ or $K[\beta][\alpha]$. You are invited to extend Proposition 6.8 to the case of any finite number of elements of the extension: Let E be a finite extension field of F, and let $\alpha_1, \ldots, \alpha_n$ be elements of E, all algebraic over F. Then $F[\alpha_1, \ldots, \alpha_n]$ is defined to be $F[\alpha_1][\ldots][\alpha_n]$, and we get the same resulting field from adjoining the a_j in every possible order. The resulting field is a finite extension of F.

Theorem 6.6 tells how to construct a field from the root of an irreducible polynomial. What if we just have the polynomial? Can we *construct* a root of it somehow? If you look at the structure of $F[\alpha]$ you will see that it depends only on the minimal polynomial. We will show how to use that polynomial to construct an extension in which the polynomial has a root.

Let F be a field, and let $f(X)$ be an irreducible element of $F[X]$. We will construct a field E that is an extension of F such that there is $\alpha \in E$ with $f(\alpha) = 0$. The precise details are abstract and they involve what looks like a "bait and switch," where we replace the field F by an isomorphic copy.

It will be convenient to use the indeterminate Y rather than X for the moment. Write the polynomial f with this indeterminate: as $f(Y)$, and consider the ideal $I = f(Y) \cdot F[Y]$ and the quotient ring $F[Y]/I$. Since f is irreducible, Theorem 3.7 shows that $f(Y)$ is a prime element of $F[Y]$, and so Corollary 3.11 shows that I is a maximal ideal of $F[Y]$, and then Proposition 2.16 shows that $F[Y]/I$ is a field, call it E. The field E will end up looking like $F[\alpha]$ for some α. To see this we need to replace F by a copy.

Define $\sigma : F \to E$ by $\sigma(b) = b + I$. It is very easy to see that σ is a ring homomorphism.[3] By Proposition 5.3, the homomorphism σ has a unique extension to a ring homomorphism from $F[X]$ to $E[X]$, where $\sigma(X) = X$. We will use σ to name this extension. Here is the actual formula: for $h(X) \in F[X]$, write

$$h(X) = \sum_{j=0}^{m} h_j \cdot X^j \quad \text{and then} \quad \sigma(h)(X) = \sum_{j=0}^{m} (h_j + I) \cdot X^j$$

Here is the heart of the matter: The polynomial $\sigma(f)(X)$, in the ring $E[X]$, has $Y + I$ as a root in E. Indeed, if the degree of $f(X)$ is n, then

$$\sigma(f)(Y + I) = \sum_{j=0}^{n} (f_j + I) \cdot (Y + I)^j = \sum_{j=0}^{n} (f_j + I) \cdot (Y^j + I)$$

$$= \sum_{j=0}^{n} f_j \cdot Y^j + I = f(Y) + I = I$$

The ideal I is the zero element of E, and so $\sigma(f)(X)$ has root $Y + I$ in E.

---

[3]Notice that this homomorphism is a restriction to F of the canonical homomorphism.

Now we transfer over to E completely. The restriction of the homomorphism σ to F is easily seen to be one to one, for if $a, b \in F$ and $\sigma(a) = \sigma(b)$, then $a + I = b + I$ leads to $a - b \in I$. Thus, $a - b$ is a multiple of the polynomial $f(X)$. Since $f(X)$ is irreducible, it is not constant, and so $a - b$ would have to be 0. Thus, $a = b$ and σ is one to one.

Since σ maps F into E one to one, we choose to identify F with its image $\sigma(F)$. In this way, F becomes a subfield of E. Under this identification, the polynomial $f(X)$ becomes the polynomial $\sigma(f)(X)$. This latter polynomial has the same degree as $f(X)$ and it is irreducible. Now $f(X)$ becomes an element of $E[X]$ and has $Y + I$ as a root in E. In class, we will discuss how to mimic this construction in such a way as to avoid *replacing* the field F with a copy.

Now that we have the existence of the adjoined field, we will go back to using X for the polynomial symbol over all fields.

Later on we will constantly face the situation of having $f(X) \in F[X]$ (not necessarily irreducible!) and needing to work with the roots of $f(X)$. This will necessitate the construction of an extension of F in which $f(X)$ has roots. To describe how to do this, we first introduce some terminology. Let $f(X) \in F[X]$, and let E be an extension of F. We say that $f(X)$ *splits* in E if it factors into $\deg(f(X))$ polynomials of degree one in $E[X]$. If $\alpha X - \beta$ is one of these factors, then $\alpha, \beta \in E$ and β/α is a root of $f(X)$. If $f(X)$ splits in E, and if $\alpha_1, \ldots, \alpha_k$ are the roots of $f(X)$ in E, then we say that $F[\alpha_1, \ldots, \alpha_k]$ is a *splitting field* for $f(X)$ over F. Recall that the order of adjoining among the α_j does not matter.

It is an easy induction argument to produce a splitting field.

THEOREM 6.9. *Let F be a field and $f(X) \in F[X]$ of degree at least one. Then there is a splitting field E for $f(X)$ over F, and any splitting field is a finite extension of F.*

PROOF. We first prove the existence of a field in which $f(X)$ splits, by induction on $\deg(f(X))$. If this degree is one, then $f(X)$ splits in F. Let $\deg(f(X))$ be greater than one. By the construction above, there is an extension K for F in which $f(X)$ has a root α. In particular, by the Factor Theorem, we can write $f(X) = (X - \alpha)g(X)$ for some $g(X) \in K[X]$. Because $\deg(g(X)) < \deg(f(X))$, induction grants us a field L containing K in which $g(X)$ splits. Since $\alpha \in L$ it is clear that $f(X)$ splits in L.

Let $f(X)$ split in L, with roots $\alpha_1, \ldots, \alpha_k$. Then $E = F[\alpha_1, \ldots, \alpha_k]$ is a splitting field for $f(X)$ over F, and E is a finite extension of F by the remarks following Proposition 6.8.   □

We can use Theorem 6.9 to construct some finite fields.

PROPOSITION 6.10. *Let p be a positive integer prime, and let n be a positive integer. Let E be a splitting field for the polynomial $X^{p^n} - X$ over $\mathbb{Z}_p$. Then E has order p^n and for all $\alpha \in E$, we have $\alpha^{p^n} = \alpha$.*

PROOF. Because E contains $\mathbb{Z}_p$, Proposition 2.5 shows that its characteristic is p. Then Proposition 2.21 shows that the function on E defined by sending x to x^p is a ring homomorphism. If σ is the composite of n copies of this function, then σ is a ring homomorphism, as well. Observe that $\sigma(\alpha) = \alpha^{p^n}$.

Define F be the set of $\alpha \in E$ such that $\sigma(\alpha) = \alpha$. We claim that F is a subring of E. Indeed, $\sigma(1) = 1$, so that $1 \in F$. And if $x, y \in F$, then

$$\sigma(x - y) = \sigma(x) - \sigma(y) = x - y \quad \text{and} \quad \sigma(xy) = \sigma(x)\sigma(y) = xy$$

Proposition 2.3 shows that F is a subring of E. Moreover, F is a field, for if $0 \neq \alpha \in F$, so that $\sigma(\alpha) = \alpha$, then $\sigma(\alpha^{-1}) = \alpha^{-1}$, so that $\alpha^{-1} \in F$.

If $\alpha \in E$ is a root of the polynomial $X^{p^n} - X$, then $\sigma(\alpha) = \alpha$, and so $\alpha \in F$. Since the polynomial $X^{p^n} - X$ splits in E, we now see that it splits in the field F. This proves that $E = F$. Now we see that $\alpha \in E$ implies that $\alpha^{p^n} = \alpha$.

Let E have m elements; to complete the proof we need to show that $m = p^n$. Let the elements of E be α_j with $1 \le j \le m$, and then there are exponents e_j such that

$$X^{p^n} - X = \prod_{j=1}^{m}(X - \alpha_j)^{e_j}$$

We will show that each e_j is 1. Indeed, you have worked out properties of the derivative of a polynomial that are relevant to this. If $(X - \alpha_j)^2$ divides $X^{p^n} - X$, then α_j is a root of the derivative of the latter polynomial. This derivative is -1. The derivative has no roots, and this contradiction completes the proof.   □

If E is a field of order p^n containing $\mathbb{Z}_p$, then $|E : \mathbb{Z}_p| = n$, since we can use a basis over $\mathbb{Z}_p$ to count the elements of E.

There is a simple corollary of Proposition 6.10 that you may have seen. For a prime p, that proposition finds a splitting field for $X^p - X$ and shows that it has order p. Then the field must be $\mathbb{Z}_p$, and so for each $\alpha \in \mathbb{Z}_p$ we have $\alpha^p = \alpha$. Reading this in the integers, we see that if $m \in \mathbb{Z}$, then $m^p \equiv m \bmod p$, and this last fact is called *Fermat's Little Theorem*.

An example might help. Suppose we want to find a field of order $27 = 3^3$. Proposition 6.10 gives you one: a splitting field for $X^{27} - X$ over $\mathbb{Z}_3$, but that description isn't very concrete. The field E we want will have $|E : \mathbb{Z}_3| = 3$, and so we might try describing E as a simple extension $E = \mathbb{Z}_3[\alpha]$ where α's minimal polynomial has degree 3. Since the minimal polynomial is irreducible, we might look for an irreducible polynomial of degree 3 over $\mathbb{Z}_3$. Here's one: $X^3 + X^2 + 2$. We can construct E from this polynomial. Let $\alpha^3 + \alpha^2 + 2 = 0$, and then $\mathbb{Z}_3[\alpha]$ will be a field of order 27.

3. Primitive Elements

Now we prove one of the most important theorems in this course. A form of this result appears in Galois' hurried notes written the night before he fought a fatal duel, and it constitutes a fundamental insight into the problem of solving polynomial equations.[4] We will prove this theorem in two cases; the first for fields that contain the rational numbers – such a field is called a *rational field*.

LEMMA 6.11. *Let F be a rational field, and let $f(X)$ be an irreducible element of $F[X]$. Let E be an extension of F in which $f(X)$ has a root α. Then $(X - \alpha)^2$ does not divide $f(X)$. In particular, $f(X)$ has $\deg(f(X))$ distinct roots in a splitting field for it.*

PROOF. The last sentence of the lemma follows from the other conclusion. If $(X - \alpha)^2$ did divide $f(X)$, then a problem on p.109 shows that $f(X)$ and $f'(X)$ would have $X - \alpha$ as a common factor. In other words, $f'(\alpha) = 0$. Proposition 6.5 then forces $f(X)$ to divide $f'(X)$.

Since $\mathbb{Q} \subseteq F$, a problem on p.109 shows that $f'(X)$ cannot be 0, and it has degree less than that of $f(X)$, Proposition 6.5 implies that $f'(X)$ is the zero polynomial, which is not the case. This contradiction proves the lemma.   □

THE PRIMITIVE ELEMENT THEOREM. *Let F be a rational field, and let E be a finite extension of F. Then there is $\delta \in E$ such that $E = F[\delta]$.*

PROOF. We use induction on $|E : F|$. If this degree is one, then $E = F = F[0]$ and we are done. Assume $|E : F| > 1$. Throughout, Proposition 6.8 will ensure that all extensions we encounter are finite extensions of F. Since we are claiming this in advance, you should be careful to check this for each field.

Claim 1: $E = F[\alpha, \beta]$ for some $\alpha, \beta \in E$.

---

[4]Galois' work was collected into three papers in French; see [**8**].

Proof. Since $|E : F| > 1$, there is $\alpha \in E$ which is not an element of F. Then $F[\alpha]$ is not F, and so $|F[\alpha] : F| > 1$. By Theorem 6.4, we have

$$|E : F| = |E : F[\alpha]| \cdot |F[\alpha] : F|$$

so that $|E : F[\alpha]| < |E : F|$. Induction allows us to apply the theorem to E as an extension of $F[\alpha]$, and we conclude that $E = F[\alpha][\beta]$ for some $\beta \in E$. The claim holds.

Now we describe how to find δ in E such that $E = F[\delta]$. Since $E = F[\alpha, \beta]$, we expect to manufacture δ from α and β. Let $f(X)$ be the minimal polynomial for α over F and $g(X)$ that for β over F. Note that $f(X)$ and $g(X)$ are elements of $E[X]$, and so Theorem 6.9 gives us a splitting field K for $f(X)g(X)$ over E. (Here is an example where you must convince yourself that K is a finite extension of F.) Let $\alpha_1, \ldots, \alpha_n$ be the distinct roots of $f(X)$ in K and $\beta_1, \ldots, \beta_m$ those of $g(X)$ in K. For convenience, let $\alpha = \alpha_1$ and $\beta = \beta_1$. Here is the key step: the set

$$S = \left\{ \frac{\alpha_i - \alpha}{\beta_j - \beta} \mid 1 \leq i \leq n, 1 < j \leq m \right\}$$

is a finite subset of K. Since $\mathbb{Q}$ is infinite, there is a rational number $r \notin S$. Note that the case $i = 1$ in the formulation of S includes 0 as an element of S, thus $r \neq 0$. Since $\mathbb{Q} \subseteq F$, we have that $r \in F$. Put $\delta = r \cdot \beta - \alpha$. For the particular claims in the rest of the proof, observe carefully which field is being considered.

Let $h(X) = g((X + \delta)/r)$. (Recall that r is not zero.) Then $h(X)$ is an element of $F[\delta][X]$. The elements $f(X)$ and $g(X)$ of $F[X]$ are also in $F[\delta][X]$. Since $f(X)$ is not zero, it has a GCD with $h(X)$, and all GCD's of $h(X)$ and $f(X)$ are associates, and therefore there is a unique one $w(X)$ which is monic. Note that $w(X) \in F[\delta][X]$.

Claim 2: $w(X) = X - \alpha$.

Proof. We will compute in K and $K[X]$. Indeed,

$$h(\alpha) = g((\alpha + \delta)/r) = g(r\beta/r) = g(\beta) = 0$$

The Factor Theorem then makes $X - \alpha$ a factor of $h(X)$ in $K[X]$. Since $f(\alpha) = 0$, we have that $X - \alpha$ is also a factor of $f(X)$, hence $X - \alpha$ is a factor of $w(X)$.

Since $w(X)$ is a GCD for $f(X)$ and $h(X)$ in $F[\delta][X]$, it divides these polynomials in $F[\delta][X]$. Therefore, $w(X)$ divides $f(X)$ and $h(X)$ in $K[X]$ as well. Since $w(X)$ divides $f(X)$, the roots of $w(X)$ are among the α_i. Let α_i be a root of $w(X)$, and then α_i is a root of $h(X)$, since $w(X)$ also divides $h(X)$. Thus $0 = h(\alpha_i) = g((\alpha_i + \delta)/r)$. We know the roots of $g(X)$, and so $(\alpha_i + \delta)/r = \beta_j$ for some j, so that $\alpha_i = r\beta_j - \delta$. Substituting the definition of δ:

$$(6.2) \qquad \alpha_i = r\beta_j - \delta = r\beta_j - (r\beta - \alpha) = \alpha + r(\beta_j - \beta)$$

If $j \neq 1$, then we can solve for r:

$$r = \frac{\alpha_i - \alpha}{\beta_j - \beta}$$

a contradiction to the choice of r. Thus $j = 1$, so that $\beta_j = \beta_1 = \beta$, and equation (6.2) now reads $\alpha_i = \alpha$.

This proves that α is the only root of $w(X)$, and it follows by unique factorization in $K[X]$ that $w(X) = (X - \alpha)^e$ for some e. But again use that $w(X)$ divides $f(X)$. Lemma 6.11 says that $(X - \alpha)^2$ does not divide $f(X)$, and now we must have that $w(X) = X - \alpha$.

Claim 3: $E = F[\delta]$.

Proof. The coefficients of $w(X)$ are elements of $F[\delta]$. Since $w(X) = X - \alpha$, the coefficient α is an element of $F[\delta]$. Then $\beta = (\delta + \alpha)/r$ is an element of $F[\delta]$, and now we have $F[\alpha, \beta] \subseteq F[\delta]$. Already $F[\delta] \subseteq E = F[\alpha, \beta]$. The claim holds and the proof is complete. $\square$

There is another case of the Primitive Element Theorem that is accessible, given our knowledge of finite fields. We will not make as much use of the following case.

PROPOSITION 6.12. *Let F be a finite field, and let E be a finite extension of F. Then there is $\alpha \in E$ such that $E = F[\alpha]$.*

PROOF. Since every element of E is a finite linear combination of elements of F with a basis, we see that E is finite. Theorem 5.9 shows that the non-zero elements of E are all powers of some given $\alpha \in E$. It is easy to see that $E = F[\alpha]$.   □

4. Problems

1. Let $F \subset E$ be fields with $|E : F| = 2$. Prove the following.

(a) If $\alpha \in E$ and $\alpha \notin F$, then $1, \alpha$ is a basis for E over F.

(b) There is an irreducible quadratic polynomial $f(X) \in F[X]$ such that $f(\alpha) = 0$.

(c) If the characteristic of F is not 2, show that there is $\beta \in E$ such that $\beta^2 \in F$ and $E = F[\beta]$.

2. Let E be an extension field of $\mathbb{Z}_2$ with $|E : \mathbb{Z}_2| = 2$. Show that if $\alpha^2 \in \mathbb{Z}_2$, then $E \neq \mathbb{Z}_2[\alpha]$.

3. Let $F \subseteq E \subseteq D$ be fields, and suppose that $|D : F|$ is prime. Show that $E = F$ or $E = D$.

4. Let p be a positive integer prime, and let n be a positive integer. Let E be a field of order p^n containing $\mathbb{Z}_p$. Show that $|E : \mathbb{Z}_p| = n$.

5. Recall the field $\mathbb{Z}_2[\alpha]$ where $\alpha^3 = \alpha + 1$. Find a basis for this field over $\mathbb{Z}_2$, and find an element of this field of multiplicative order 7.

6. Let $\alpha = 1 + \sqrt[4]{7}$. Express $1/\alpha$ as a rational combination of powers of $\sqrt[4]{7}$.

7. Find a field of order 16 and an element of multiplicative order 15. (Hint: you want $\mathbb{Z}_2 \subseteq E$ with $|E : \mathbb{Z}_2| = 4$; look for an irreducible element of $\mathbb{Z}_2[X]$ of degree 4.)

8. Find a field E with $\mathbb{Q} \subset E \subset \mathbb{C}$ that is a splitting field for $x^4 - 2$. Compute $|E : \mathbb{Q}|$ and give a basis for E over $\mathbb{Q}$.

9. Let p be a positive integer prime, and let n and m be positive integers. Let E be a field of order p^n. Show that E has a subfield of order p^m if and only if m divides n. (Hint: For one direction, use Theorem 6.4. For the other direction, if $\alpha^{p^m} = \alpha$, use induction to show that $\alpha^{p^{qm}} = \alpha$ for $q = 1, 2, \dots$.)

10. Let $f(X) \in \mathbb{Q}[X]$ be irreducible, of degree 3, and with exactly one real root. Show that a splitting field E for $f(X)$ over $\mathbb{Q}$ satisfies $|E : \mathbb{Q}| = 6$.

11. Find $f(X) \in \mathbb{Q}[X]$ of degree 3 with a real root α such that $\mathbb{Q}[\alpha]$ is a splitting field for $f(X)$ over $\mathbb{Q}$.

12. An *algebraic integer* is a complex number that is a root of some non-zero, monic polynomial with integer coefficients.

(a) If $\alpha \in \mathbb{C}$ is algebraic over the rationals, then there is a positive integer n such that $n \cdot \alpha$ is an algebraic integer.

(b) If $r \in \mathbb{Q}$ is an algebraic integer, then $r \in \mathbb{Z}$.

13. Let $\alpha = i \cdot \sqrt{3}$ (in the complex numbers). Let a, b be odd integers. Show that $(a + b \cdot \alpha)/2$ is an algebraic integer. (Hint: show that $a^2 + 3b^2$ is divisible by 4.)

14. Let $z = \exp(2\pi i/5)$. In each case, write the given element in the form $f(z)$ where $f(X) \in \mathbb{Q}[X]$ has degree less than $|\mathbb{Q}[z] : \mathbb{Q}|$.

        a) $1/z$ **b)** $(1+z)^{-2}$ **c)** z^{123}

15. Let $w = \exp(2\pi i/3)$. Show that $\mathbb{Q}[w]$ is the unique subfield $E \subseteq \mathbb{Q}(w, \sqrt[3]{2})$ such that $|E : \mathbb{Q}| = 2$. (Hint: what is $|E[w] : E|$?)

16. Let $\alpha = \exp(2\pi i/7)$. Define $\beta = \alpha + \alpha^2 + \alpha^4$. Show that β is the root of a non-zero quadratic polynomial with integer coefficients. (Notice that $\{1, 2, 4\}$ is a subgroup, under multiplication, of U_7.) Can you find a similar element δ that is a sum of powers of $\exp(2\pi i/11)$ and such that δ is a root of a non-zero quadratic polynomial over with nteger coefficients?

17. Let p and q be distinct integer primes. (Note: cases such as $p = 3$, $q = -3$ are allowed.)

(a) Prove that $\sqrt{p} \notin \mathbb{Q}[\sqrt{q}]$.
(b) Prove that $X^2 - p$ is irreducible over $\mathbb{Q}[\sqrt{q}]$.
(c) Conclude that $|\mathbb{Q}[\sqrt{p}, \sqrt{q}] : \mathbb{Q}| = 4$.
(d) Find $\alpha \in \mathbb{Q}[\sqrt{p}, \sqrt{q}]$ such that $\mathbb{Q}[\alpha] = \mathbb{Q}[\sqrt{p}, \sqrt{q}]$.

18. Find $\delta \in \mathbb{C}$ so that $\mathbb{Q}[\delta]$ is a splitting field for $X^3 - 2$.

19. Find $\alpha \in \mathbb{C}$ so that $\mathbb{Q}[\alpha]$ is a splitting field for $(X^2 - 2) \cdot (X^3 - 2)$.

20. Let F be a field and let $f(X) = X^2 + b \cdot X + c$ be irreducible in $F[X]$. Let α be a root of $f(X)$ in some field. Define

$$A = \begin{pmatrix} 0 & 1 \\ -c & -b \end{pmatrix}$$

Show that $\sigma(r + s \cdot \alpha) = r \cdot I_2 + s \cdot A$ defines a one to one ring homomorphism to $\mathfrak{M}(2, F)$. (Thus, the image of σ is isomorphic to $F[\alpha]$; this gives another way to construct $F[\alpha]$ in the first place.)

21. Continue the notation of the previous problem, and let R be the image of σ. Let $B \in \mathfrak{M}(2, F)$. Show that B commutes with every element of R if and only if $B \in R$. (Hint: show that B commutes with every element of R if and only if it commutes with A.)

CHAPTER 7

The Galois Group of an Extension

The Galois group[1] can be defined very generally. We think it will help at first to work mainly with rational fields, nodding occasionally to the finite fields.

1. Automorphisms

We begin with a lemma concerning homomorphisms.

PROPOSITION 7.1. *Let E and L be extensions of the field F, and let σ be a ring homomorphism from E to L such that $\sigma(r) = r$ for all $r \in F$. Let $f(X) \in F[X]$, and let $\alpha \in E$. Then $\sigma(f(\alpha)) = f(\sigma(\alpha))$. In particular, if α is a root of $f(X)$, then $\sigma(\alpha)$ is also a root of $f(X)$.*

PROOF. This is an easy, direct calculation. Assume the notation of the assertion, let $n = \deg(f(X))$, and compute

$$f(\sigma(\alpha)) = \sum_{j=0}^{n} f_j \cdot \sigma(\alpha)^j = \sum_{j=0}^{n} \sigma(f_j) \cdot \sigma(\alpha)^j$$

$$= \sum_{j=0}^{n} \sigma(f_j \cdot \alpha^j) = \sigma\left(\sum_{j=0}^{n} f_j \cdot \alpha^j\right) = \sigma(f(\alpha))$$

$\square$

---

[1]We have previously referenced Galois' work [**8**].

135

Let E be a finite extension of the field F. The *Galois group* of E *over* F is defined to be the set of all field isomorphisms $\sigma : E \to E$ such that $\sigma(r) = r$ for all $r \in F$. This Galois group is denoted $\mathrm{Gal}(E/F)$.

The equation $\sigma(r) = r$ for all $r \in F$, required for $\sigma \in \mathrm{Gal}(E/F)$, can be described by saying that σ *fixes* the elements of F. An isomorphism from a field to itself is called an *automorphism*, and so $\mathrm{Gal}(E/F)$ is the set of *automorphisms of E that fix F*.

It is straightforward that $\mathrm{Gal}(E/F)$ is a group under function composition. (For instance, an automorphism has an inverse that is, itself, an automorphism.) Here is how the automorphisms come about.

THEOREM 7.2. *Let F be a field, E a finite extension field, and suppose that $E = F[\alpha]$ for some $\alpha \in E$. Let $f(X)$ be the minimal polynomial for α over F. Then the function that sends $\sigma \in Gal(E/F)$ to the element $\sigma(\alpha)$ of E is a one to one, onto function from $Gal(E/F)$ to the set of roots of $f(X)$ in E.*

PROOF. If $\sigma \in \mathrm{Gal}(E/F)$, then Proposition 7.1 says that $\sigma(\alpha)$ is a root of $f(X)$, and so the function described, from $\mathrm{Gal}(E/F)$ to the roots of $f(X)$ in E, is defined.

We show that this function is one to one: let $\sigma, \tau \in \mathrm{Gal}(E/F)$ with $\sigma(\alpha) = \tau(\alpha)$. Then $(\tau^{-1}\sigma)(\alpha) = \alpha$. Let $\beta \in E$, and then the definition of $F[\alpha]$ allows us to write $\beta = r(\alpha)$ where $r(X) \in F[X]$. Proposition 7.1 then computes

$$(\tau^{-1}\sigma)(\beta) = (\tau^{-1}\sigma)(r(\alpha))$$
$$= r((\tau^{-1}\sigma)(\alpha)) = r(\alpha) = \beta$$

This shows that $\tau^{-1} \cdot \sigma$ is the identity function on E (the identity element of $\mathrm{Gal}(E/F)$), so that $\tau = \sigma$. This establishes that the function is one to one from $\mathrm{Gal}(E/F)$ to the roots.

Onto? Let β be a root of $f(X)$ in E. By Theorem 6.7 there is an isomorphism σ from $F[\alpha]$ to $F[\beta]$ such that $\sigma(\alpha) = \beta$ and with $\sigma(r) = r$ for all $r \in F$. We have $\sigma \in \mathrm{Gal}(E/F)$ if we show that $F[\beta] = E$. Anyway, $f(X)$ is a monic irreducible element of $F[X]$ with β as a root, and then Proposition 6.5 shows that $f(X)$ is the minimal polynomial of β over F. The relevance of this is that $|F[\beta] : F| = \deg(f(X))$. But then

$$|E : F| = |F[\alpha] : F| = \deg(f(X))$$

and so $|E : F| = |F[\beta] : F|$. Using Theorem 6.4 we get

$$|E : F| = |E : F[\beta]| \cdot |F[\beta] : F| = |E : F[\beta]| \cdot |E : F|$$

It follows that $|E : F[\beta]| = 1$, and this proves that $E = F[\beta]$, as needed. □

Theorem 7.2 gives an algorithm for determining the Galois group of a simple extension. You should immediately work examples! Some are given in the exercises at the end.

Theorem 7.2 leads easily to the following.

COROLLARY 7.3. *Let E be a finite extension of the rational field F. Then $|Gal(E/F)| \leq |E : F|$.*

PROOF. The Primitive Element Theorem shows that $E = F[\alpha]$ for some $\alpha \in E$. The number $|E : F|$ is the degree of the minimal polynomial $f(X)$ for α over F, where $E = F[\alpha]$, and so $|E : F| = \deg(f(X))$.

Corollary 5.8 says that $f(X)$ has less than or equal to $\deg(f(X))$ roots in E. By Theorem 7.2, the number of roots of $f(X)$ in E is the order of $\mathrm{Gal}(E/F)$. □

It is usually a good question when an inequality is an equality. In this subject, the answer is a goldmine. We will say that E is *Galois* over F if E

is a finite extension of F with $|\mathrm{Gal}(E/F)| = |E : F|$. What does Theorem 7.2 say about this? Let $E = F[\alpha]$, and let $f(X)$ be the minimal polynomial for α over F. Then E is Galois over F if and only if $f(X)$ has $\deg(f(X)) = |E : F|$ roots in E.

THEOREM 7.4. *Let E be a finite extension of the rational field F with $E = F[\alpha]$ and $f(X)$ the minimal polynomial for α over F. Then E is Galois over F if and only if E is a splitting field for $f(X)$ over F, and if and only if $f(X)$ splits in E.*

PROOF. If E is Galois over F, then by Theorem 7.2 the field E contains $\deg(f(X))$ roots of $f(X)$. Thus $f(X)$ splits in E. Since $E = F[\alpha]$, we certainly can obtain E from F by adjoining all the roots of $f(X)$, and thus E is a splitting field for $f(X)$.

If E is a splitting field for $f(X)$, then Lemma 6.11 shows that E contains $\deg(f(X))$ roots of $f(X)$, and so E is Galois over F by Theorem 7.2. $\qquad\square$

The field E of Theorem 7.4 is the splitting field of an *irreducible* polynomial. We now show that the splitting field for *every* polynomial over a rational field is Galois.

THEOREM 7.5. *Let F be a rational field, let $f(X) \in F[X]$, and let E be a splitting field for $f(X)$ over F. Then E is Galois over F.*

PROOF. By Theorem 6.9, the field E is a finite extension of F. By the Primitive Element Theorem, there is $\alpha \in E$ such that $E = F[\alpha]$. Let $g(X) \in F[X]$ be the minimal polynomial for α over F. We will show that $g(X)$ splits in E, and then Theorem 7.4 will show that E is Galois over F.

Let K be a splitting field for $g(X)$ over E, so that $g(X)$ splits in K. Let $\beta \in K$ be a root of $g(X)$. We intend to use Theorem 6.7: the fields $F[\alpha]$ and $F[\beta]$ are extensions of F, and $g(X)$ is irreducible in $F[X]$, having α, β as roots. By Theorem 6.7, there is a field isomorphism $\sigma : F[\alpha] \to F[\beta]$ such that $\sigma(r) = r$ for all $r \in F$, and with $\sigma(\alpha) = \beta$.

Now we go back to the polynomial $f(X)$. If $\gamma \in K$ is a root of $f(X)$ then by Proposition 7.1, we have $\sigma(\gamma)$ is a root of $f(X)$ in K. Since $f(X)$ splits in $E \subseteq K$, we see that $\sigma(\gamma) \in E$. The field $E = F[\alpha]$ is generated by F and the roots of $f(X)$. Since σ maps F to itself and maps the roots of $f(X)$ among themselves, we see that σ maps $F[\alpha]$ into itself. In particular $\beta = \sigma(\alpha) \in F[\alpha] = E$.

We have shown that the roots of $g(X)$ in K are in E. Since K was constructed to be a splitting field for $g(X)$ over E, we conclude that $K = E$. $\square$

One consequence of Theorem 7.5 is that splitting fields for a given polynomial over a given field are isomorphic, as are the Galois groups involved.

THEOREM 7.6. *Let F be a rational field and $f(X) \in F[X]$. If E and J are splitting fields for $f(X)$ over F then*

(a) there is an isomorphism σ from E to J which fixes F;
(b) the groups $Gal(E/F)$ and $Gal(J/F)$ are isomorphic.

PROOF. The Primitive Element Theorem finds $\alpha \in E$ such that $E = F[\alpha]$. Let $g(X)$ be the minimal polynomial for α in $F[X]$. Let K be a splitting field for $g(X)$ over J, and let β be a root of $g(X)$ in K.

By Theorem 6.7 there is a field isomorphism $\sigma : F[\alpha] \to F[\beta]$ fixing F and with $\sigma(\alpha) = \beta$. By Proposition 7.1, σ sends roots of $f(X)$ in $E = F[\alpha]$ to roots of $f(X)$ in K, and these roots are in J. Since E is generated by F and the roots in E of $f(X)$, we see that $\sigma(E)$ is generated by F and the roots in J of $f(X)$, and it follows that $\sigma(E) \subseteq J$. In particular, $\beta \in J$, so that $F[\beta] \subseteq J$.

We have

$$|E : F| = |F[\alpha] : F| = \deg(g(X)) = |F[\beta] : F| \leq |J : F|$$

Thus, $|E : F| \leq |J : F|$. Switching the roles of E, J in the same argument, we find that $|J : F| \leq |E : F|$, and so $|E : F| = |J : F|$. It follows that $J = F[\beta]$, and so σ is an isomorphism from E to J, fixing F. This proves (a).

For (b), given $\tau \in \mathrm{Gal}(E/F)$ define the function $\sigma\tau\sigma^{-1}$ from J to J. It is trivial that this function is an automorphism of J that fixes F and so $\sigma\tau\sigma^{-1} \in \mathrm{Gal}(J/F)$. You will observe that the mapping sending τ to $\sigma\tau\sigma^{-1}$ is a group isomorphism from $\mathrm{Gal}(E/F)$ to $\mathrm{Gal}(J/F)$. $\qquad\square$

2. The Subgroup Theorem

We are now ready to unveil a correspondence between the subgroups of the Galois group of an extension and subfields of that extension. This correspondence has many remarkable consequences. Let E be a finite extension of F (not necessarily Galois), and let H be a subgroup of $\mathrm{Gal}(E/F)$. Define $\mathrm{Fix}(H)$ to be the set of $\alpha \in E$ such that $\sigma(\alpha) = \alpha$ for all $\sigma \in H$. This is the set of all field elements *fixed* by all the elements of H; it is called the *fixed field* of H.

Certainly F is contained in $\mathrm{Fix}(H)$ since every element of $\mathrm{Gal}(E/F)$ fixes all of F. Because the elements of H are automorphisms, $\mathrm{Fix}(H)$ is in fact a field. Thus $\mathrm{Fix}(H)$ is an extension of F contained in E. Here is the first miracle in this subject.

LEMMA 7.7. *Let E be a finite extension of the rational field F, and let H be a subgroup of $\mathrm{Gal}(E/F)$. Then $|H| = |E : \mathrm{Fix}(H)|$, the field E is Galois over $\mathrm{Fix}(H)$, and $H = \mathrm{Gal}(E/\mathrm{Fix}(H))$.*

PROOF. Put $K = \mathrm{Fix}(H)$, and view E as an extension field of K. Corollary 7.3 says that $|\mathrm{Gal}(E/K)| \leq |E : K|$. Now H is a group of automorphisms of E which fix K, and so H is a subgroup of $\mathrm{Gal}(E/K)$. It follows that

$$(7.1) \qquad\qquad |H| \leq |\mathrm{Gal}(E/K)| \leq |E : K|$$

Now we will get a reverse inequality. Use the Primitive Element Theorem to write $E = K[\alpha]$ for some $\alpha \in E$. We define

$$f(X) = \prod_{\sigma \in H} (X - \sigma(\alpha))$$

We claim that $f(X) \in K[X]$. Before proving this, let us show that it will complete the proof. We have $\deg(f(X)) = |H|$. If $g(X)$ is the minimal polynomial for α over K, then $f(\alpha) = 0$ forces that $g(X)$ divides $f(X)$, and in particular that $\deg(g) \leq \deg(f)$. We know that $\deg(g) = |K[\alpha] : K| = |E : K|$, and we conclude that

$$|H| = \deg(f) \geq |E : K|$$

In light of this, the inequalities of equation (7.1) must be equalities! Thus $H = \mathrm{Gal}(E/K)$, and $|H| = |E : K|$. This establishes the conclusions of the lemma.

We are left to show that $f(X) \in K[X]$. We do this using the definition of K. Let $\tau \in H$, so that τ is an isomorphism from E to E. By Proposition 5.3 we can extend τ to an isomorphism from $E[X]$ to $E[X]$ with $\tau(X) = X$. Applying this extended τ to $f(X)$ we see that

$$\tau(f(X)) = \prod_{\sigma \in H} \tau(X - \sigma(\alpha)) = \prod_{\sigma \in H} (X - \tau\sigma(\alpha))$$

Since H is a group, this last polynomial is $f(X)$. Thus τ fixes $f(X)$, and therefore τ fixes the coefficients of $f(X)$. Since this last statement is true for all elements of H, we have that the coefficients of $f(X)$ are elements of $\mathrm{Fix}(H)$, which is K. $\qquad\square$

Let us apply Lemma 7.7 to $H = \mathrm{Gal}(E/F)$. It says that E is always a Galois extension of $\mathrm{Fix}(\mathrm{Gal}(E/F))$. If $E = F[\alpha]$, and $f(X)$ is the minimal polynomial of α over $\mathrm{Fix}(\mathrm{Gal}(E/F))$, then Theorem 7.4 says that $f(X)$ splits in E.

Recall that when H is a subgroup of the finite group G, Lagrange's Theorem on p.10 tells us that the number $|G : H|$ of left cosets of H in G multiplied by the order $|H|$ of H gives the order of G.

We can now establish the promised correspondence between fields and subgroups. This is the main theorem of this chapter.

GALOIS' CORRESPONDENCE. *Let E be a Galois extension of the rational field F, and put $G = Gal(E/F)$. The function Fix is a one to one correspondence between the set of subgroups of G and the set of fields between E and F.*

(a) *Let H be a subgroup of G, and put $K = Fix(H)$. Then E is Galois over K with $H = Gal(E/K)$, with $|H| = |E : K|$, and with $|G : H| = |K : F|$.*

(b) *Let K be a field between E and F, and put $H = Gal(E/K)$. Then H is a subgroup of G, and $K = Fix(H)$.*

PROOF. Let H be a subgroup of G, and let $K = \text{Fix}(H)$. The conclusions of (a) follow from Lemma 7.7 except for the last equality. Since E is Galois over F, we have $|G| = |E : F|$. Then

$$|G : H| \cdot |H| = |G| = |E|$$
$$= |E : K| \cdot |K : F|$$
$$= |H| \cdot |K : F|$$

and this proves that $|G : H| = |K : F|$. The statement (a) holds.

In the notation of (b), let $E = F[\alpha]$, and let $f(X)$ be the minimal polynomial for α over F. By Theorem 7.4 we have that E is a splitting field for $f(X)$ over F. Then it follows easily that E is a splitting field for $f(X)$ over K, and therefore E is Galois over K by Theorem 7.5.

The definition of Galois group shows that H is a subgroup of G. We then know that $|E : \text{Fix}(H)| = |H|$, courtesy of Lemma 7.7. Also, $|H| = |E : K|$.

Since $K \subseteq \mathrm{Fix}(H)$, it follows that $K = \mathrm{Fix}(H)$, and (b) holds.

That Fix is one to one and onto is immediate. It is one to one for if H and J are subgroups of G such that $\mathrm{Fix}(H) = \mathrm{Fix}(J)$, then (a) leads to $H = \mathrm{Gal}(E/\mathrm{Fix}(H))$ and $J = \mathrm{Gal}(E/\mathrm{Fix}(J))$, so that $H = J$. As for being onto, if K is a field between E and F, and if $H = \mathrm{Gal}(E/K)$, then (b) shows that $K = \mathrm{Fix}(H)$, so that K is in the image of Fix. $\qquad \square$

As an application, we prove a result which will be used to determine which regular n-gons are constructible using compass and straightedge. This topic will be taken up in Chapter 8, Section 2

PROPOSITION 7.8. *Suppose that p is a prime and $p - 1$ is a power of 2. Let $z = \exp(2\pi i/p)$. Then $X^p - 1$ splits in $\mathbb{Q}[z]$ and there are fields*

$$\mathbb{Q} = F_0 \subseteq F_1 \subseteq \ldots \subseteq F_e = \mathbb{Q}[z]$$

and $|F_j : F_{j-1}| = 2$, for $1 \leq j \leq e$.

PROOF. Since the roots of $X^p - 1$ are powers of z, we see that the polynomial splits in $\mathbb{Q}[z]$. In particular, $\mathbb{Q}[z]$ is Galois over $\mathbb{Q}$.

By Theorem 5.16, the minimal polynomial for z is $\Phi_p(X)$ and it has roots z^k where $1 \leq k \leq p - 1$. Theorem 7.2 shows how to construct the Galois group: there are $p - 1$ elements σ_k of that Galois group where $\sigma_k(z) = z^k$. You should show that these automorphisms commute, so that the Galois group is abelian.

Since G is abelian, Proposition 1.4 says that if m is a positive integer dividing $|G|$, then G has a subgroup of order m. We will use this fact to construct some subgroups of $\mathrm{Gal}(\mathbb{Q}[z]/\mathbb{Q})$.

First, define $H_0 = \mathrm{Gal}(\mathbb{Q}[z]/\mathbb{Q})$. Since $|H_0| = 2^e$, this abelian group has a subgroup H_1 of order 2^{e-1}. The abelian group H_1 has a subgroup H_2 of order

2^{e-2}. Continuing in this way we obtain subgroups $|H_j|$ of order 2^{e-j} with

$$\mathrm{Gal}(\mathbb{Q}[z]/\mathbb{Q}) = H_0 \supseteq H_1 \supseteq H_2 \supseteq \ldots \supseteq H_e = \{\, 1 \,\}$$

and $|H_j : H_{j+1}| = 2$ for $0 \leq j < e$.

Define $F_j = \mathrm{Fix}(H_j)$ for $0 \leq j \leq e$, then the Galois Correspondence shows that

$$\mathbb{Q} = F_0 \subseteq F_1 \subseteq \ldots \subseteq F_e = \mathbb{Q}[z]$$

and $|F_j : F_{j-1}| = 2$, for $1 \leq j \leq e$. $\square$

3. Normal Extensions

This section is short, but it gives the setting for the original discovery by Galois of normal subgroups of groups. If E is a Galois extension of F, and if K is a field contained in E and containing F, then the Galois Correspondence shows that E is always Galois over K. It is natural to ask when K is Galois over F; we know that the answer is, "Not always!" This very natural question leads to the discovery of the normal subgroup.

It will be convenient to collect an elaborate hypothesis.

Hypothesis 0. Let F be a rational field. Let E be a Galois extension of F with $G = \mathrm{Gal}(E/F)$. Let $H \subseteq G$ and $K = \mathrm{Fix}(H)$. Suppose that $K = F[\alpha]$. Let $f(X)$ be the minimal polynomial for α over F.

PROPOSITION 7.9. *Assume Hypothesis 0. If $\sigma \in G$, then $\sigma(\alpha)$ is a root of $f(X)$ in E. For $\sigma, \tau \in G$, $\sigma(\alpha) = \tau(\alpha)$ if and only if the two left cosets $\sigma \cdot H$ and $\tau \cdot H$ are equal. We have that $f(X)$ splits in E. If β is a root of $f(X)$ in E, then there is $\sigma \in G$ such that $\beta = \sigma(\alpha)$.*

PROOF. By Proposition 7.1 $\sigma(\alpha)$ is a root of $f(X)$ for all $\sigma \in G$. If $\sigma(\alpha) = \tau(\alpha)$, then $\sigma^{-1}\tau(\alpha) = \alpha$, and it follows from Proposition 7.1 that $\sigma^{-1}\tau$ fixes $F(\alpha) = K$. In other words, $\sigma^{-1}\tau \in H$, which says that $\sigma \cdot H = \tau \cdot H$.

Conversely, if $\sigma \cdot H = \tau \cdot H$, then $\sigma = \tau \cdot \delta$ for some $\delta \in H$, so that $\sigma(\alpha) = \tau\delta(\alpha) = \tau(\alpha)$, since H fixes α. This proves the if and only if assertion.

We see that the distinct cosets of H in G count distinct roots of $f(X)$ in E. Thus $f(X)$ has at least $|G : H|$ roots in E. On the other hand the Galois Correspondence shows that

$$|K : F| = |G : H| \quad \text{and} \quad |K : F| = |F[\alpha] : F| = \deg(f(X))$$

Therefore, $f(X)$ has $\deg(f(X))$ roots in E; it must split in E!

The polynomial $g(X) = \prod_{\sigma \in G}(X - \sigma(\alpha))$ is fixed by every element of G, and so by the Galois Correspondence, $g(X) \in F[X]$. Clearly $g(X)$ has α as a root, and since $f(X)$ is the minimal polynomial for α over F, we see that $f(X)$ divides $g(X)$. This implies that every root β of $f(X)$ has the form $\sigma(\alpha)$ for some $\sigma \in G$. $\qquad\square$

We pause to consider a profound consequence of Proposition 7.9.

COROLLARY 7.10. *Let E be a splitting field for the element $g(X)$ of $F[X]$. Let $f(X)$ be an irreducible element of $F[X]$ which has a root in E. Then $f(X)$ splits in E.*

PROOF. By Theorem 7.5, E is Galois over F. Let α be a root of $f(X)$ in E, and put $K = F[\alpha]$. Then by Proposition 7.9, $f(X)$ splits in E. $\qquad\square$

In other words, the fact that one polynomial splits in E has implications for a possibly large set of other polynomials. Notice that the statement of this corollary makes no mention of Galois groups; this is a result purely about fields. It is likely that such a result would have remained unknown long after Galois, but for his theory.

Assume Hypothesis 0. We ask again, when is K Galois over F? Theorem 7.4 says that K is Galois exactly when $f(X)$ splits in K. Proposition 7.9 says that $f(X)$ splits in E, and it identifies the roots $\sigma(\alpha)$ for various $\sigma \in G$.

Let us ask when these roots are in K. How do you know $\beta \in E$ is in K? Since $K = \text{Fix}(H)$, we have $\beta \in K$ if and only if $\delta(\beta) = \beta$ for all $\delta \in H$. For $\delta \in H$, the equation $\delta\sigma(\alpha) = \sigma(\alpha)$ is equivalent to $\sigma^{-1}\delta\sigma(\alpha) = \alpha$. The element $\sigma^{-1}\delta\sigma$ fixes α if and only if it fixes K. The automorphism $\sigma^{-1}\delta\sigma$ of E fixes K if and only if it lies in $\text{Gal}(E/K)$. The group $\text{Gal}(E/K)$ is H, by the Galois Correspondence. Thus $\sigma(\alpha) \in K$ if and only if $\sigma^{-1}\delta\sigma \in H$ for all $\delta \in H$ and $\sigma \in G$. Recall Proposition 1.3 of Chapter 1: H has the property in question if and only if it is a normal subgroup of G.

THE NORMAL SUBGROUP THEOREM. *Assume Hypothesis 0. Then K is Galois over F if and only if H is a normal subgroup of G. If H is normal in G, then $\text{Gal}(K/F)$ is isomorphic to G/H.*

PROOF. The first conclusion follows from the paragraph before this theorem. Assume that K is Galois over F, and let $\sigma \in G$. Then by the previous argument $\sigma(\alpha)$ is an element of K, and so σ maps K to K. Let σ' be the restriction of σ to K, then σ' is an automorphism of K which fixes F (an element of $\text{Gal}(K/F)$). The mapping from $G = \text{Gal}(E/F)$ to $\text{Gal}(K/F)$ sending σ to σ' is easily seen to be a group homomorphism. The kernel of this homomorphism is the set of elements of G which act like the identity function on K, and these are the elements which fix K, in other words the elements of H! It follows that the image of this homomorphism is isomorphic to G/H. In particular, this image has order $|G/H| = |G : H|$, and we already know by the Galois Correspondence that $|G/H| = |K : F|$. This last number is also $|\text{Gal}(K/F)|$, and so our homomorphism must be onto. The proof is complete. $\square$

4. Problems

1. Find the following Galois groups by describing what the automorphisms do to the element that generates each extension. In (d), we have $w = \exp(2\pi i/8)$.

    a) $\mathrm{Gal}(\mathbb{Q}[i]/\mathbb{Q})$ **b)** $\mathrm{Gal}(\mathbb{Q}[\sqrt[4]{2}]/\mathbb{Q})$

    c) $\mathrm{Gal}(\mathbb{Q}[i \cdot \sqrt[4]{3}]/\mathbb{Q}[\sqrt{3}])$ **d)** $\mathrm{Gal}(\mathbb{Q}[w]/\mathbb{Q})$

2. This problem gives a commonly used technique for finding a Galois group. Let $w = \exp(2\pi i/3)$, and $E = \mathbb{Q}[\sqrt[3]{2}, w]$. It is easy to see that E is a splitting field for $X^3 - 2$ over $\mathbb{Q}$. We will find $\mathrm{Gal}(E/\mathbb{Q})$ as follows. Let $F = \mathbb{Q}[\sqrt[3]{2}]$.

(a) Show that $|F : \mathbb{Q}| = 3$ and $|E : F| = 2$, so that $|E : \mathbb{Q}| = 6$.

(b) Show that $\mathrm{Gal}(E/F) = \langle \sigma \rangle$ where σ has order 2.

(c) Show that $\mathrm{Gal}(E/\mathbb{Q}[w]) = \langle \alpha \rangle$ where α has order 3.

(d) Show that $\sigma, \sigma \cdot \alpha, \sigma \cdot \alpha^2, \alpha, \alpha^2$ are all distinct. (Determine what each does to w and to $\sqrt[3]{2}$. These are the non-identity elements of $\mathrm{Gal}(E/\mathbb{Q})$.)

3. Describe $\mathrm{Gal}(E/\mathbb{Q})$ where E is a splitting field for $X^4 - 2$.

4. Describe $\mathrm{Gal}(E/\mathbb{Q})$ where E is a splitting field for $X^8 - 1$ over $\mathbb{Q}$.

5. Suppose that F is a rational field, that E is an extension field, and that $|E : F| = 2$. Show that E is Galois over F.

6. Find $\alpha \in \mathbb{C}$ such that $\mathbb{Q}[\alpha]$ is Galois over $\mathbb{Q}$ and $|\mathbb{Q}[\alpha] : \mathbb{Q}| = 3$.

7. Let F be a rational field, and let $f(X) \in F[X]$ be irreducible of degree n. Let E be a splitting field for $f(X)$ over F, and let R be the set of roots of $f(X)$ in E. For $\alpha \in \mathrm{Gal}(E/F)$, define $P(\alpha) : R \to R$ by $P(\alpha)(r) = \alpha(r)$ for all $r \in R$.

(a) Show that the image of $P(\alpha)$ is in R.

(b) Show that P is a one to one group homomorphism from $\mathrm{Gal}(E/F)$ to $\mathbb{S}_R$.

(c) Conclude that $|E : F|$ divides $n!$.

8. Give the Galois correspondence explicitly for the splitting field $E \subseteq \mathbb{C}$ for $X^4 - 2$ over the rationals.

9. Give the Galois correspondence explicitly for the splitting field $K \subseteq \mathbb{C}$ $X^7 - 1$.

10. Suppose that F is a rational field, E is a Galois extension field, and $\mathrm{Gal}(E/F)$ is abelian. Let $f(X)$ be an irreducible element of $F[X]$ and suppose that $f(X)$ has a root α in E. Then $f(X)$ splits in $F[\alpha]$.

11. Let $E \subseteq \mathbb{C}$ be a splitting field for $X^4 - 2$ over the rationals. The Galois group G has order 8 and an element σ of order 4.

(a) Show that $\langle \sigma \rangle$ is normal in G.
(b) Describe the fixed field of σ, and show that it is Galois over the rationals.
(c) Show that $\langle \sigma^2 \rangle$ is normal in G.
(d) Describe the fixed field of σ^2 and show that it is Galois over the rationals.

12. Let $K \subseteq \mathbb{C}$ be a splitting field for $X^7 - 1$. There is a subfield L of K with $|L : \mathbb{Q}| = 3$. Find $\delta \in E$ with $\mathbb{Q}[\delta] = L$ and find the roots in L of the minimal polynomial for δ over $\mathbb{Q}$.

13. Let $\mathbb{Z}_p \subset E$ be a finite field of order p^n where p is a prime and n is a positive integer. We know that $\sigma : E \to E$ with $\sigma(b) = b^p$ is an automorphism. Show that σ has order n. Note: It follows that E is Galois over $\mathbb{Z}_p$ and $\langle \sigma \rangle = \mathrm{Gal}(E/\mathbb{Z}_p)$.

14. Let F be a rational field, let n be a positive integer, and let E be a splitting field for $X^n - 1$ over F. Show that $\mathrm{Gal}(E/F)$ is abelian. (Hint: the roots form a finite group in E, and so Theorem 5.9 shows that the group has the form $\langle \alpha \rangle$. Show that each element of $\mathrm{Gal}(E/F)$ is determined by what it does to α.)

CHAPTER 8

Applications

1. The Cyclotomic Polynomials

We need three facts about the group U_n of units in $\mathbb{Z}_n$. Each of our facts is fairly elementary and we will not bother to prove them here; proofs are given in many books introducing abstract algebra – for instance, [**11**].

(a) The integer m is in U_n if and only if the GCD of n, m is 1.

(b) If $\langle x \rangle$ has order n, then $\langle x \rangle = \langle x^m \rangle$ if and only if $m \in U_n$.

(c) If $n = p_1^{e_1} \cdots p_k^{e_k}$ for distinct positive integer primes p_j and positive integers e_j, then

$$|U_n| = \prod_{j=1}^{k} p_j^{e_j - 1} \cdot (p_j - 1)$$

Given a positive integer n, we want to find the minimal polynomials, over the rationals, of the complex n-th roots of 1. From Chapter 1, we know that the n-th roots of 1 are the powers of $w = \exp(2\pi i/n)$. We index these powers using $\mathbb{Z}_n$, since if $a, b \in \mathbb{Z}$, then $w^a = w^b$ if and only if $a \equiv b \bmod n$. Fact (b) above shows that the complex roots of 1 of multiplicative order n are precisely w^m where $m \in U_n$. We define

$$\Phi_n(X) = \prod_{m \in U_n} (X - w^m)$$

The polynomial $\Phi_n(X)$ is called the n-th *cyclotomic polynomial*. The roots of $\Phi_n(X)$ are the complex numbers of order n.

We will prove that the cyclotomic polynomials have integer coefficients and that they account for the factorization of $X^n - 1$ into irreducible elements of $\mathbb{Q}[X]$.

THEOREM 8.1. *Let n be a positive integer. Then*

(a) The polynomial $\Phi_n(X)$ is monic and has integer coefficients.
(b) $\Phi_n(X)$ is irreducible in $\mathbb{Q}[X]$.

PROOF. Statement (a) is proved by induction on n; observe that $\Phi_1(X) = X - 1$. Suppose that the statement is true for all positive integers n with $n < k$. Let $D[k]$ be the set of positive integer divisors of k, except for k itself, and define

$$g(X) = \prod_{n \in D[k]} \Phi_n(X)$$

Observe that

$$X^k - 1 = \Phi_k(X) \cdot g(X)$$

since, as remarked above, each root of $X^k - 1$ has order dividing k. By induction, $g(X)$ is a monic element of $\mathbb{Z}[X]$.

We claim that $\Phi_k(X) \in \mathbb{Q}[X]$. Indeed, Proposition 5.5 finds $q(X), r(X) \in \mathbb{Q}[X]$ such that

$$X^k - 1 = q(X) \cdot g(X) + r(X) \quad \text{and} \quad \deg(r(X)) < \deg(g(X))$$

Each complex root of $g(X)$ is a root of $X^k - 1$, and so it is a root of $r(X)$. Since the roots of $g(X)$ are not repeated, this contradicts that $\deg(r(X)) < \deg(g(X))$ unless $r(X) = 0$. We see that $q(X) = \Phi_n(X)$, and this proves that $\Phi_n(X) \in \mathbb{Q}[X]$.

Gauss' Lemma finds a rational number r such that $r \cdot \Phi_n(X)$ and $g(X)/r$ have integer coefficients. Since $\Phi_x(X)$ and $g(X)$ are monic, $r = \pm 1$. We conclude that $\Phi_n(X) \in \mathbb{Z}[X]$, and statement (a) is proved.

For (b), suppose that $\Phi_n(X)$ is not irreducible. Let $f(X)$ be a monic irreducible factor of $\Phi_n(X)$ in $\mathbb{Q}[X]$. By Gauss' Lemma, we have $f(X) \in \mathbb{Z}[X]$.

Claim. There is a root α of $f(X)$ and a prime p not dividing n such that α^p is not a root of $f(X)$.

We consider pairs (β, k) where β is a root of $f(X)$ and k is a positive integer such that β^k is a root of $\Phi_n(X)$ that is **not** a root of $f(X)$. We claim that such pairs exist: Indeed, let β a root of $f(X)$. Since β has order n, all the roots of $X^n - 1$ are powers of β, and so each root of $\Phi_n(X)$ is such a power. Since $f(X)$ is not $\Phi_n(X)$, there is a root δ of $\Phi_n(X)$ that is not a root of $f(X)$. Thus, $\delta = \beta^k$ for a positive integer $k > 1$. Then (β, k) is one of our pairs.

Choose a pair (α, p) with p minimal, and we claim that p is prime. Since $p > 1$, it has a positive integer prime divisor r. Write $p = rq$, so that $1 \leq q < p$. We know that p, n have GCD 1, and therefore q, n have GCD 1, and so α^q is a root of $\Phi_n(X)$. If α^q is not a root of $f(X)$, then (α, q) is a pair with $q < p$, a contradiction. Thus, α^q is a root of f, and so (α^q, r) is a pair. This contradicts the minimality of p unless $p = r$, as claimed.

We focus on the pair (α, p) where p is prime. Write $X^n - 1 = f(X) \cdot g(X)$ where $g(X) \in \mathbb{Z}[X]$. We see that α^p is a root of $X^n - 1$ but not a root of $f(X)$; thus, α^p is a root of $g(X)$. This shows that α is a root of $g(X^p)$. It follows that $f(X)$ divides $g(X^p)$, since $f(X)$ is the minimal polynomial of α.

The canonical homomorphism $\sigma : \mathbb{Z} \to \mathbb{Z}_p$ extends to $\sigma : \mathbb{Z}[X] \to \mathbb{Z}_p[X]$ where $\sigma(X) = X$. We have $\sigma(X^n - 1) = \sigma(f(X)) \cdot \sigma(g(X))$. Also $\sigma(f(X))$ divides $\sigma(g(X^p)) = \sigma(g(X))^p$.

Let E be a splitting field for $\sigma(X^n - 1)$ over $\mathbb{Z}_p$, and let β be a root of $\sigma(f(X))$. We see that β is a root of $\sigma(g(X))^p$, and so β is a root of $\sigma(g(X))$. It follows that β is a repeated root of $\sigma(X^n - 1)$. The derivative of $\sigma(X^n - 1)$ is $n * X^{n-1}$; since p does not divide n, this derivative does not have a root in common with $\sigma(X^n - 1)$, a contradiction. $\qquad \square$

Let $z = \exp(2\pi i/n)$, and then $\mathbb{Q}[z]$ is a splitting field for $X^n - 1$, and so it is a Galois extension of the rationals. The fact that $\Phi_n(X)$ is irreducible tells us about $Gal(\mathbb{Q}[z]/\mathbb{Q})$.

THEOREM 8.2. *Let $z = \exp(2\pi i/n)$ for some positive integer n. There is a group isomorphism $f : U_n \to Gal(\mathbb{Q}[z]/\mathbb{Q})$, where if $m \in U_n$, then $f(m)(z) = z^m$.*

PROOF. If $m \in U_n$, then z^m has multiplicative order n, and so z^m is a root of the minimal polynomial $\Phi_n(X)$ of z. There is therefore an element $f(m)$ of the Galois group with $f(m)(z) = z^m$. Conversely, all the roots of $\Phi_n(X)$ have the form z^m for $m \in U_n$.

The function f is one to one and onto. To see that it is a group homomorphism (and therefore an isomorphism) we have only to compute for units m, k:

$$f(m)(f(k)(z)) = f(m)(z^k) = \left[f(m)(z) \right]^k$$
$$= \left[z^m \right]^k = z^{mk} = f(mk)(z)$$

Since $f(mk)$ and $f(m) \cdot f(k)$ (composite function!) agree at z and fix $\mathbb{Q}$, they are equal.   □

2. Geometric Constructions

Which geometric figures may be constructed in the plane using only a compass (for drawing circles) and a straightedge? This question goes back at least to Euclid, but it was Gauss who first gave the answer in a satisfactory way.[1] Remarkably, he was eighteen at the time.

We begin with some definitions. Let S be a set of points in the plane. Define $L(S)$ to be the set of lines through pairs of elements of S. Define $C(S)$

---

[1]Gauss' proof of what amounts to our Theorem 8.4 and Theorem 8.6 and occurs in Section VII of [**9**].

to be the set of all circles having center in S and passing through a point of S. A point R in the plane is *constructed* from S if there are distinct figures A and B in $L(S) \cup C(S)$ such that R is an intersection point of A and B.

Let S be a set of points in the plane, and suppose that $F \subseteq \mathbb{C}$ is a field such that for all $(a, b) \in S$ we have $a, b \in F$. Let (x, y) be constructed from S. Gauss observed the remarkable fact that $|F[x, y] : F| \leq 2$! We will observe this for ourselves by considering the three possible ways that (x, y) is constructed from S.

Case 1: (x, y) is the intersection of two elements of $L(S)$.

Consider one of the lines of intersection. By definition of $L(S)$ there are distinct points (r, s) and (t, u) in S such that this line passes through these points. The reader knows that $(s - u)(X - r) + (t - r)(Y - s) = 0$ is an equation of this line, and that this equation can be written in the form $aX + bY = c$ where $a = (s - u)$, and $b = (t - r)$, and $c = r(s - u) + s(t - r)$. These last three formulas lead from $r, s, t, u \in F$ to $a, b, c \in F$. A similar equation $dX + eY = f$ can be found for the other line.

Since the two lines are distinct and intersect, the system of equations

$$aX + bY = c \quad \text{and} \quad dX + eY = f$$

has a unique solution, and the point of solution is (x, y) by the definition of this point. You know from linear algebra that we can write x and y in terms of the coefficients a, b, c, d, e, f. In other words, $x, y \in F$! Certainly, then $|F[x, y] : F| = 1 \leq 2$.

Case 2: (x, y) is an intersection of an element of $L(S)$ and $C(S)$.

Let the line be through (r, s) and (t, u) and the circle with center at (e, f) and through (g, h). The circle is

$$(X - e)^2 + (Y - f)^2 = (e - g)^2 + (f - h)^2$$

and the line is as before $aX + bY = c$ where $a, b, c \in F$.

As in Case 1, the point (x, y) is a solution to both equations. If $b \neq 0$, then we can solve the equation of the line for y in terms of a, b, c, x. Substitution for y into the equation of the circle then shows that x is a root of a quadratic polynomial with coefficients in F. It follows that $|F[x] : F| \leq 2$. Furthermore, the formula for y in terms of x shows that $y \in F[x]$ so that $F[x] = F[x, y]$ and then $|F[x, y] : F| \leq 2$.

The previous paragraph depended on $b \neq 0$. If $b = 0$, then the line is vertical $X = c/a$, so that $x = c/a \in F$. Then, in the equation for the circle, all the unknowns are in F except possibly Y. This shows that y is a root of a quadratic polynomial over F, so that $|F[y] : F| \leq 2$. This time $x \in F$ implies that $F[y] = F[x, y]$ and once again $|F[x, y] : F| \leq 2$.

Case 3: (x, y) is an intersection of two elements of $C(S)$.

Suppose that the two circles are

$$(X - a)^2 + (Y - b)^2 = (a - c)^2 + (b - d)^2$$
$$(X - e)^2 + (Y - f)^2 = (e - g)^2 + (f - h)^2$$

where $a, b, c, d, e, f, g, h \in F$. As before, (x, y) is a simultaneous solution. Let $u = x - a$ and $v = y - b$, so then the equations of the circles yield

$$u^2 + v^2 = r^2$$
$$(u + a - e)^2 + (v + b - f)^2 = s^2$$

where $r, s \in F$. Expanding out the squares, subtracting the first equation from the second, and collecting expressions in a, b, e, f, r, s we obtain

(8.1) $2u(a - e) + 2v(b - f) = t$ where $t \in F$

Because the two circles have to be distinct yet intersecting (in (x, y)!), they have different centers. In other words, either $a - e \neq 0$ or $b - f \neq 0$. In any case, we can solve (8.1) for u or for v, and then substitution into one of the other equations for u, v gives a quadratic equation in the remaining unknown.

This proves that $|F[u, v] : F| \leq 2$. The definitions of u and of v show that $F[u, v] = F[x, y]$, so now $|F[x, y] : F| \leq 2$

We summarize the three cases.

PROPOSITION 8.3. *Let S be a set of points in the plane, and let $F \subseteq \mathbb{C}$ be a field such that $(a, b) \in S$ implies that $a, b \in F$. Let (x, y) be constructed from S. Then $|F[x, y] : F| \leq 2$.*

Now we are ready to define the geometric construction of a point. Let (x, y) be a point in the plane. Then (x, y) is *constructible* if there is a sequence of points $a_0, a_1, \ldots, a_n$ where

   (a) $a_0 = (0, 0)$ and $a_1 = (1, 0)$
   (b) for each $j \geq 1$, the point a_{j+1} is constructed from $\{a_0, a_1, \ldots, a_j\}$
   (c) $a_n = (x, y)$

The idea of condition (a) is that we are allowed to start by drawing $(0, 0)$ and $(1, 0)$ anywhere we wish. After that, we draw lines and circles through previously drawn points and collect the (new) points of intersection one at a time. We allow an arbitrary but finite number of steps. If S is the set of all constructible points, then $L(S)$ and $C(S)$ are, respectively, the set of *constructible lines* and *constructible circles*.

The point (x, y) can be identified with the complex number $x + i \cdot y$. We say that the complex number is *constructible* if and only if the point is constructible. Proposition 8.3 shows that the set of constructible complex numbers is severely limited. The following theorem is due to Gauss.

THEOREM 8.4. *Let z be a complex number. If z is constructible, then it is algebraic over $\mathbb{Q}$ and $|\mathbb{Q}[z] : \mathbb{Q}|$ is a power of 2. Moreover, there are fields*

$$\mathbb{Q} = F_0 \subseteq F_1 \subseteq \ldots \subseteq F_n \subseteq \mathbb{C}$$

such that

(a) $z \in F_n$,

(b) $|F_j : F_{j-1}| \leq 2$, for $1 \leq j \leq n$.

PROOF. Assume that z is constructible, and write $a_0, a_1, \ldots, a_n = z$ as in the definition of constructible.

Define $F_0 = \mathbb{Q}$, and define $F_1 = \mathbb{Q}[i]$, and notice that $|F_1 : F_0| = 2$. We will define fields F_j for which $|F_{j+1} : F_j| \leq 2$ and such that if $k \leq j$, then F_j will contain the X and Y coordinates of a_k. We see that F_0 contains $(0,0) = 0$ and F_1 contains the point $(0,1) = i$.

Working by induction, assume that F_j has been defined for some $j \geq 1$ with the required properties. Write $a_{j+1} = x + iy$. Since, by definition, a_{j+1} is constructed from $a_0, \ldots, a_j$, and since F_j is a subfield of $\mathbb{C}$ containing the X and Y coordinates of these points, Proposition 8.3 shows us that $|F_j[x,y] : F_j| \leq 2$. We define $F_{j+1} = F_j[x,y]$ and observe that F_{j+1} has the required properties.

We end up with a sequence of fields with the correct degrees. Write $z = x + iy$, and then $x, y \in F_n$. Since $i \in F_1 \subseteq F_n$ we see that $i \in F_n$ so now $z = x + iy \in F_n$.

Furthermore, that $|F_{j+1} : F_j| \leq 2$ shows us that $|F_n : \mathbb{Q}|$ is a power of 2. It follows that z is algebraic over $\mathbb{Q}$ and since $\mathbb{Q}[z]$ is a subfield of F_n we have that $|\mathbb{Q}[z] : \mathbb{Q}|$ divides $|F_n : \mathbb{Q}|$, so that $|\mathbb{Q}[z] : \mathbb{Q}|$ is a power of 2.   □

As a technicality, we mention that it is possible for $|\mathbb{Q}[z] : \mathbb{Q}|$ to be a power of 2 *without* z being constructible (without the presence of intermediate fields). For example, there is a rational polynomial of degree 4 whose roots are not constructible.[2]

Theorem 8.4 answers some questions of classical geometry. Is the cube root of a constructible number constructible? If so then $\sqrt[3]{2}$ is constructible.

---

[2]We will not give a specific example except to say that it suffices to arrange it so that the Galois group of the polynomial over the rationals is the alternating group $\mathbb{A}_4$. This group has a subgroup K with $|A : K| = 4$. The fixed field of K gives the required elements.

But $\sqrt[3]{2}$ is a root of the irreducible polynomial $X^3 - 2$ over $\mathbb{Q}$, and so $|\mathbb{Q}[\sqrt[3]{2}]$:
$\mathbb{Q}| = 3$. Since 3 is not a power of 2, the number $\sqrt[3]{2}$ is not constructible! In
particular, there can be no general construction of cube roots.

If α is the measure of an angle resulting from a construction, then, as we
will show momentarily, $\cos(\alpha)$ is a constructible real number. Taking this fact
as a given, we ask, "Is there a general technique for trisecting angles?" One
can easily show that a 3,4,5 right triangle is constructible, and so the angle
α with $\cos(\alpha) = 3/5$ is constructible. If the angle α can be trisected, then
$\cos(\alpha/3)$ is constructible. The cosine addition formula shows that

$$4\cos^3(\alpha/3) - 3\cos(\alpha/3) = \cos(\alpha) = 3/5$$

so that

$$20\cos^3(\alpha/3) - 15\cos(\alpha/3) - 3 = 0$$

Eisenstein's Criteria ($p = 3$) show that $20X^3 - 15X - 3$ is irreducible, and
so $|\mathbb{Q}[\cos(\alpha/3)] : \mathbb{Q}| = 3$. This angle cannot be trisected! There can be no
general construction of angle trisections.

Is π constructible? This one escaped even Gauss. In 1882, Lindemann
proved that π is not algebraic over the rationals, and so Theorem 8.4 shows
that π is not constructible. All known proofs of this result are extremely
difficult.

Another famous problem: "Which regular polygons are constructible?"
For a positive integer n, the unit regular n-gon is the polygon with n equal
length sides inscribed in the unit circle, passing through the point $(1,0)$. It is
easy to see that the polygon meets the unit circle at the n-th roots of 1, and
so we say that the regular n-gon is constructible if and only if the n-th roots
of 1 are constructible. Theorem 8.4 gives a partial, negative answer to this
question.

THEOREM 8.5. *Let n be a positive integer, and suppose that a regular n-gon is constructible. Then $n = 2^e \cdot p_1 \cdots p_k$, where e is a non-negative integer, and the p_j are distinct positive odd integer primes such that $p_j - 1$ is a power of 2.*

PROOF. Let $z = \exp(2\pi i/n)$, and Theorem 8.1 shows that

$$|\mathbb{Q}[z] : \mathbb{Q}| = \deg(\Phi_n(X)) = |U_n|$$

If z is constructible, then Theorem 8.4 shows that $|\mathbb{Q}[z] : \mathbb{Q}|$ is a power of 2. By the formula for $|U_n|$, the result follows.   □

For instance, we cannot construct a regular 7-gon, 9-gon, 11-gon, 13-gon. Notice, however that $17 - 1 = 2^4$. Is a regular 17-gon constructible? Gauss proved the converse of Theorem 8.5 to answer this question in the affirmative. His proof gives an explicit series of compass-straightedge steps for the construction. This was a remarkable feat, representing the discovery of a really new fact of plane geometry two thousand years after the work of the (very clever) Greek geometers.

To get at the converse of Theorem 8.5, we need to establish some algebra for constructible numbers. Here is a list of facts that can be demonstrated in order – the first two come from Euclid [7]. In the following, capital letters stand for points already constructed. The segment AB will be identified with its length when convenient.

(1) (Euclid Proposition 1.) Given A, B distinct, can construct C such that $\triangle ABC$ is equilateral.
(2) (Euclid Proposition 2.) Given A and distinct B, C, can construct D such that $AD = BC$.
(3) Given A, B distinct, can construct D such that $\angle ABD$ is a right angle. For instance, can construct at least one point on the y-axis not the origin.

(4) Given A not on the line through B, C, can construct D on the line with $\angle BDA$ a right angle. (Unless $\angle ABC$ is already a right angle.)

(5) Can construct $(b, 0)$ if and only if can construct $(0, b)$. For instance, can construct i.

(6) (Euclid Proposition 23.) Given $\angle ABC$ and distinct points D, E can construct F so that $\angle ABC = \angle DEF$.

(7) Given (a, b), can construct $(a, 0)$ and $(b, 0)$.

(8) Given $(a, 0)$ and $(b, 0)$, can construct (a, b). In particular, if θ is a constructible angle, then $(\cos(\theta), 0)$ is constructible.

(9) Given A, B can construct $A + B$.

(10) Given $(a, 0)$ and $(b, 0)$, can construct $(ab, 0)$.

(11) Given A and B, can construct $A \cdot B$.

(12) Given $(a, 0)$ and $(b, 0)$ with $b \neq 0$, can construct $(a/b, 0)$.

(13) Given A, B with $B \neq 0$, can construct A/B.

(14) Given $(a, 0)$ with $a > 0$, can construct $(\sqrt{a}, 0)$.

(15) Given $\angle ABC$ can construct D such that $\angle ABD$ bisects the angle.

(16) Given A, can construct B such that $B^2 = A$.

THEOREM 8.6. *Suppose there are fields*

$$\mathbb{Q} = F_0 \subseteq F_1 \subseteq \ldots \subseteq F_n \subseteq \mathbb{C}$$

such that $|F_j : F_{j-1}| \leq 2$, for $1 \leq j \leq n$. Then every element of F_n is constructible.

PROOF. Induction on n shows that the elements of $F = F_{n-1}$ are constructible. If $z \in F_n$, then $|F[z] : F| \leq 2$. If $|F[z] : F| = 1$, then $z \in F$ is constructible, and so we can assume that z is the root of a quadratic polynomial $X^2 + b \cdot X + c$ where $b, c \in F$. Then, of course, z is one of the two numbers $\frac{1}{2} \cdot (-b \pm \alpha)$ where $\alpha \in \mathbb{C}$ satisfies $\alpha^2 = b^2 - 4c$.

By the construction steps above, we see that α is constructible, and then z is constructible. $\square$

And the converse of Theorem 8.5.

THEOREM 8.7. *Suppose that $n = n = 2^e \cdot p_1 \cdots p_k$, where e is a non-negative integer, and the p_j are distinct odd integer primes such that $p_j - 1$ is a power of 2. Then the regular n-gon is constructible.*

PROOF. For each j, let $z_j = \exp(2\pi i/p_j)$, and Proposition 7.8 shows that Theorem 8.6 applies to $\mathbb{Q}[z_j]$. Thus, z_j is constructible.

It follows from our constructions that $z = z_1 \cdots z_k$ is constructible. This complex number has order $p = p_1 \cdots p_k$. The definition of $\Phi_p(X)$ shows that all its roots are powers of z; in particular $\exp(2\pi i/p)$ is a power of z, and we see that $\exp(2\pi i/p)$ is constructible.

By repeatedly bisecting angles $2\pi/p$, we obtain the construction of the number $\exp(2\pi i/n)$. $\square$

A final word of caution. Many have misunderstood what is meant by the "impossibility of a construction." The literature is replete with algorithms that trisect angles *approximately* or that trisect *only some* angles or that change the rules of construction somehow to make trisection possible or that construct approximations to π. Mathematical theorems are precise.

3. Radical Extensions

We are ready to solve the problem that Galois addressed as the principal object of his theory. In order to understand the solution, we need to know what is meant by a *solvable group*. This term is defined in Chapter 1 on p.16. However, Galois almost certainly discovered the definition from working with fields – it would be possible to deduce the definition from the work on which we are about to engage.

We need a few definitions. Let E be a finite extension of the rational field F. Then E is an *n-radical extension* if n is a positive integer and $E = F[\alpha]$ where $\alpha^n \in F$. We first note that if E is an n-radical extension and if n divides the positive integer m, then E is also an m-radical extension, for $\alpha^n \in F$ implies $\alpha^m \in F$ in this case. You have shown that if $|E : F| = 2$, then E is a 2-radical extension of F.

On the notation: when we have $\alpha^n = \beta$, we can say that α is an n-th root of β as long as we remember that there may be several other n-th roots as well. We will not use the function notation $\sqrt[n]{\beta}$ because of this ambiguity.

We wish to study fields constructed from F by repeatedly adjoining n-th roots. Specifically, E is a *repeated radical extension* of F if there is a sequence of fields $F = E_1 \subseteq E_2 \subseteq \ldots \subseteq E_m = E$ such that for $1 < j \leq m$, the field E_j is an n_j-radical extension of E_{j-1} for some positive integer n_j. Let n be the product of the n_j, and then the relevant remark two paragraphs ago shows that each E_j is an n-radical extension of E_{j-1}. We refer to E as a repeated n-radical extension of F (getting rid of the n_j). It is clear that if K is a repeated radical extension of E, and E is a repeated radical extension of F, then K is a repeated radical extension of F.

If $f(X)$ is an irreducible element of $F[X]$, then $f(X)$ is *solvable in radicals* if there is a repeated radical extension of F in which $f(X)$ has a root. You might be interested in a field in which $f(X)$ splits; it turns out that if $f(X)$ is solvable in radicals, then there is a repeated radical extension of $f(X)$ in which $f(X)$ splits – this will fall out of a proof given below.

The following is universally regarded as one of the most beautiful and interesting theorems ever. As we mentioned above, we can hold off on solvable groups if we wish, and let the proof show us what they have to be.

GALOIS' SOLVABILITY THEOREM. *Let F be a rational field and $f(X)$ an irreducible element of $F[X]$. Let K be a splitting field for $f(X)$ over F. Then $f(X)$ is solvable in radicals if and only if $\mathrm{Gal}(K/F)$ is a solvable group. If $f(X)$ is solvable in radicals, then K is contained in a repeated radical extension of F.*

More is true: when $f(X)$ is solvable in radicals, there is a systematic way to use $\mathrm{Gal}(K/F)$ to write down the roots of $f(X)$ in terms of elements of F and n-th roots. This method, which will be evident in our proof, gives an algebraic formula for the roots of $f(X)$.

Galois' Solvability Theorem represents a triumph of the abstract method; the Galois group encapsulates the information that is needed to tell whether the polynomial is solvable. The proof uses almost everything we have done in the course!

We will prove the two implications of Galois' Solvability Theorem separately. First we work under the hypothesis that $f(X)$ is solvable in radicals. (Here we will see what the structure of the Galois group is; there is where the definition of solvability comes from.) We need to construct a field L that is in general larger than a splitting field; we will then verify that $\mathrm{Gal}(L/F)$ is solvable. Here is the construction; it is quite technical.

Assume that E is a repeated n-radical extension of F with

$$F = E_1 \subseteq \ldots \subseteq E_m = E$$

and $E_j = E_{j-1}[\alpha_j]$ with $\alpha_j^n \in E_{j-1}$ for $1 < j \leq m$. Let $g_j(X)$ be the minimal polynomial for α_j over F, for each j, and define $g(X)$ to be the product of $X^n - 1$ along with all the $g_j(X)$.

Now let L be a splitting field for $g(X)$ over E. The field L is large enough in which to do all our work; we now need some subfields. Notice that since $g(X)$ splits in L, all the $g_j(X)$ split there too. Thus we can define

(1) the field L_1 is the splitting field for $X^n - 1$ over F in L.

(2) for $j = 2, 3, \ldots, m$ define L_j to be the splitting field for $g_j(X)$ over L_{j-1} in L.

LEMMA 8.8. *Under these circumstances, for $1 \leq j \leq m$:*

(a) $E_j \subseteq L_j$

(b) L_j is Galois over F

(c) L_j is L_{j-1} adjoined by elements whose n-th powers are in L_{j-1}.

PROOF. We use induction on j. For the case $j = 1$ notice that $E_1 = F \subseteq L_1$ by definition, that L_1 is Galois over F since it is a splitting field over F, and that L_1 is generated by some n-th root of 1. Thus (a)-(c) hold for $j = 1$.

Assume we have the conclusions for $j = k - 1$. Since L_k is, by definition, a splitting field for $g_k(X)$, we have that the root α_k of $g_k(X)$ is in L_k. Also by definition the field L_k contains the field L_{k-1}. Thus $L_{k-1}[\alpha_k] \subseteq L_k$. By induction we have $E_{k-1} \subseteq L_{k-1}$ so now

$$E_k = E_{k-1}[\alpha_k] \subseteq L_{k-1}[\alpha_k] \subseteq L_k$$

which establishes (a) in the case $j = k$.

By induction L_{k-1} is Galois over F. Then L_{k-1} is a splitting field for some polynomial $h(X) \in F[X]$. We claim that L_k is a splitting field for the polynomial $h(X)g_k(X)$ over F. This will prove (b). Indeed, the definition of L_k shows that $g_k(X)$ splits in it, and that $L_{k-1} \subseteq L_k$. Thus $h(X)g_k(X)$ certainly splits in L_k. We know that adjoining the roots of $h(X)$ to F gives L_{k-1} and that adjoining the roots of $g_k(X)$ to L_{k-1} gives L_k. Therefore adjoining the roots of $h(X)g_k(X)$ to F gives L_k and now it is clear that L_k is a splitting field over F. This proves (b) in the case $j = k$.

Finally we are left with (c). Recall that $\alpha_k^n \in E_{k-1}$ so that conclusion (b) shows that $\alpha_k^n \in L_{k-1}$. Since L_k is a splitting field for $g_k(X)$, if β is a root of

$g_k(X)$, then there is $\sigma \in \text{Gal}(L_k/F)$ such that $\beta = \sigma(\alpha_k)$. Then $\beta^n = \sigma(\alpha_k^n)$. Since $\alpha_k^n \in L_{j-1}$ and L_{k-1} is Galois over F, we see that $\beta^n \in L_{j-1}$.

Then since

$$L_k = L_{k-1}[\text{all the roots of } g_k(X)]$$

we now see that L_k is gotten by adjoining n-th roots of elements of L_{k-1}, and now we see that (c) holds. Whew. $\square$

Solvable groups are composed of abelian quotient groups. The next fact shows where the abelian groups come from.

LEMMA 8.9. *Let n be as above. Let S be a field in which $X^n - 1$ splits. Let R be a Galois extension of S such that $R = S[\beta_1, \ldots, \beta_k]$ where β_j^n are nonzero elements of S for all j. Then $\text{Gal}(R/S)$ is an abelian group.*

PROOF. Let σ and τ be elements of $\text{Gal}(R/S)$. What can we say about $\sigma(\beta_j)$? Notice that β_j is a root of the polynomial $X^n - \beta_j^n$ and by hypothesis this polynomial is in $S[X]$. Thus σ must take β_j to another root of this polynomial. This polynomial has exactly n roots: $u \cdot \beta$ where u is one of the n roots of the polynomial $X^n - 1$ (recall the hypothesis that this polynomial splits in S). We see that $\sigma(\beta_j) = u \cdot \beta_j$ for some $u \in S$.

Similarly, $\tau(\beta_j) = v \cdot \beta_j$ for some $v \in S$. Thus

$$\tau(\sigma(\beta_j)) = \tau(u\beta_j) = u \cdot \tau(\beta_j)$$
$$= u \cdot v \cdot \beta_j = v \cdot u \cdot \beta_j = \sigma(\tau(\beta_j))$$

This proves that $\tau \cdot \sigma = \sigma \cdot \tau$, and so $\text{Gal}(R/S)$ is abelian. $\square$

We can now prove half of Galois' Solvability Criteria.

LEMMA 8.10. *Assume the hypothesis before and including Lemma 8.8. Let K be a splitting field for $f(X)$ over F. Then $\text{Gal}(K/F)$ is solvable.*

PROOF. We know that $E = E_m \subseteq L_m$ contains a root of $f(X)$. Corollary 7.10 then forces $f(X)$ to split in L_m, so that there is a splitting field J for $f(X)$ in L_m. By Theorem 7.6 the groups $\mathrm{Gal}(K/F)$ and $\mathrm{Gal}(J/F)$ are isomorphic. The definition of solvable shows that if $\mathrm{Gal}(J/F)$ is solvable, then $\mathrm{Gal}(K/F)$ is solvable as well.

Lemma 8.8 tells us that L_m is Galois over F. The Normal Subgroup Theorem then says that $\mathrm{Gal}(J/F)$ is isomorphic to a quotient group of $\mathrm{Gal}(L_m/F)$. Indeed, we have $\mathrm{Gal}(L_m/J)$ normal in $\mathrm{Gal}(L_m/F)$ and $\mathrm{Gal}(J/F)$ isomorphic to

$$\mathrm{Gal}(L_m/F)/\mathrm{Gal}(L_m/J)$$

Proposition 1.7 says that if $\mathrm{Gal}(L_m/F)$ is solvable, then $\mathrm{Gal}(J/F)$ is solvable.

Lemma 8.8 along with the definition of the L_j show us that Lemma 21.3 applies to L_j as an extension of L_{j-1} for $j > 1$ and we see that $\mathrm{Gal}(L_j/L_{j-1})$ is abelian. A problem on p.148 shows that L_1 is a Galois extension of F with $\mathrm{Gal}(L_1/F)$ abelian. For convenience of notation, put $L_0 = F$ and now we have that $\mathrm{Gal}(L_j/L_{j-1})$ is abelian for $i > 0$.

The rest of the proof is largely a matter of notation. We will quote the Normal Subgroup Theorem. Let $N_j = \mathrm{Gal}(L_m/L_j)$ and observe that

$$\mathrm{Gal}(L_m/L_0) = N_0 \supseteq N_1 \supseteq \ldots \supseteq N_m = \{1\}$$

and that N_{j-1}/N_j is isomorphic to $\mathrm{Gal}(L_j/L_{j-1})$. These last groups (for each j) are abelian, and the definition of solvable shows that $\mathrm{Gal}(L_m/L_0) = \mathrm{Gal}(L_m/F)$ is solvable. $\square$

4. A Non-Solvable Polynomial

You will need to remember the cycle notation for permutations, and the fact that if S, T are finite sets of same size, then the permutation groups $\mathbb{S}_S$ and $\mathbb{S}_T$ are isomorphic.

The polynomial we consider will generate a field which is Galois over the rationals, and the Galois group will be isomorphic to a subgroup of $\mathbb{S}_p$ where p is a prime. We will need to conclude that the Galois group is isomorphic to the whole group $\mathbb{S}_p$. The following lemma holds the technical fact we will need to draw the required conclusion.

LEMMA 8.11. *Assume that T is a set of prime order p. Let H be a subgroup of $\mathbb{S}_T$ such that p divides $|H|$ and such that H contains a 2-cycle. Then $H = \mathbb{S}_T$.*

PROOF. By the remark that $\mathbb{S}_T$ and $\mathbb{S}_p$ are isomorphic, we may as well assume that H is a subgroup of $\mathbb{S}_p$. We will show that H contains all the 2-cycles. Since every element of $\mathbb{S}_p$ is a product of 2-cycles, the proof will be complete.[3]

Claim. Without loss of generality, H contains the cycles (12) and $(123\ldots p)$.

Indeed, let y be the 2-cycle mentioned in the hypothesis of the Lemma. By re-numbering the points (this amounts to an induced isomorphism as explained above), we can assume that $y = (12)$.

Since p divides $|H|$, Cauchy's Theorem produces $x \in H$ such that the order of x is p. The only cycle that has order p is a p-cycle, and so x must include a p-cycle. Since there are only p points total, we see that x is a p-cycle.

The fact that x is a p-cycle on the p points shows that some power of x, say x^k, sends 1 to 2. Then $x^k \in H$ and x^k cannot be the identity permutation. Because p is prime, it follows that x^k has order p so that x^k is a p-cycle. In cycle notation $x^k = (12a_3a_4\ldots a_p)$. By re-numbering the points $3, 4, \ldots, p$, we can get $x^k = (12\ldots p)$ as desired. This proves the Claim.

We are assuming that $y = (123\ldots p) \in H$ and that $(12) \in H$.

---

[3]Every element of $\mathbb{S}_p$ is a product of cycles, and every cycle is a product of 2-cycles. See [11].

Claim. The cycles (12), (23), (34), $\ldots$, $(p\text{-}1\ p)$ are in H.

The proof is that the named cycles are none other than

$$(12), \quad y(12)y^{-1}, \quad y^2(12)y^{-2}, \quad \ldots \quad ,y^{p-2}(12)y^{2-p}$$

Claim. All the 2-cycles are in H.

Suppose that the 2-cycle is (ij) with $i < j$. We will prove that $(ij) \in H$ by induction on $j - i$. If $j - i = 1$, then (ij) is one of the 2-cycles referred to in the previous claim. These 2-cycles are in H.

Suppose that $j - i > 1$ and that cycles (ab) with $a < b$ and $b - a < j - i$ are cycles in H. Then

$$(ij) = (j\text{-}1\ j) \cdot (i\ j\text{-}1) \cdot (j\text{-}1; j)$$

This formula shows that $(ij) \in H$, as needed. $\square$

Here we link an assumption about the roots of a polynomial to the hypothesis of Lemma 8.11.

LEMMA 8.12. *Let $f(X) \in \mathbb{Q}[X]$ be irreducible of prime degree p, and assume that $f(X)$ has exactly $p - 2$ real roots. Let E be a splitting field for $f(X)$ over $\mathbb{Q}$ with $E \subseteq \mathbb{C}$. Then E is Galois over $\mathbb{Q}$ and $\mathrm{Gal}(E/\mathbb{Q})$ is isomorphic to $\mathbb{S}_p$.*

PROOF. Let $\sigma \in \mathrm{Gal}(E/\mathbb{Q})$, and observe for $\alpha \in E$ that $f(\alpha) = 0$ if and only if $f(\sigma(\alpha)) = 0$ (a fact we have used many times). In other words, if T is the set of roots of $f(X)$ in E, then σ maps T to T. We write $\bar{\sigma}$ for the restriction of σ to T.

Because σ is one to one on E, it is one to one on T. Since T is finite, σ is then onto T, so that $\bar{\sigma}$ is a permutation of T.

The function $\bar{\sigma}$ from $\mathrm{Gal}(E/\mathbb{Q})$ obtained in this way is easily seen to be a group homomorphism. Since E is generated by the roots of $f(X)$, $\bar{\sigma}$ is one to one. We use the name H for the image in $\mathbb{S}_T$ of the homomorphism $\bar{\sigma}$, and then $\mathrm{Gal}(E/\mathbb{Q})$ is isomorphic to H.

Because $f(X) \in \mathbb{Q}[X]$, it does not have repeated roots, so that $|T| = \deg(f(X)) = p$, which is prime, by hypothesis. Thus H is a subgroup of $\mathbb{S}_T$ where $|T|$ is the prime p.

Now it is time to notice that complex conjugation is a ring automorphism of $\mathbb{C}$; this is clear from Proposition 1.1. It follows, since E is a splitting field inside $\mathbb{C}$, that conjugation maps E to E. We call δ the restriction of conjugation to E, and then δ is an automorphism of E. Clearly, δ fixes $\mathbb{Q}$, so now $\delta \in \mathrm{Gal}(E/\mathbb{Q})$.

The $p-2$ real roots of $f(X)$ are fixed by δ, and it cannot fix the other two roots (since they are not real). Thus $\bar{\delta}$ is a 2-cycle, so that H contains a 2-cycle. We will now show that p divides $|H|$, and we will see that Lemma 8.11 applies to H.

If α is one of the roots of $f(X)$ in E, then $\mathbb{Q}[\alpha] \subseteq E$ and therefore $|\mathbb{Q}[\alpha] : \mathbb{Q}|$ divides $|E : \mathbb{Q}|$. But that $f(X)$ is irreducible shows that it is the minimal polynomial for α over $\mathbb{Q}$, and therefore $|\mathbb{Q}[\alpha] : \mathbb{Q}| = \deg(f) = p$. Now p divides $|E : \mathbb{Q}| = |\mathrm{Gal}(E/\mathbb{Q})| = |H|$, as needed.

Lemma 8.11 now says that H is $\mathbb{S}_T$, so that since $\mathrm{Gal}(E/\mathbb{Q})$ is isomorphic to H we see that $\mathrm{Gal}(E/\mathbb{Q})$ is isomorphic to $\mathbb{S}_p$. $\square$

Next, we need to know that the permutation groups are usually not solvable.

LEMMA 8.13. *Let n be a positive integer with $n \geq 5$. Then $\mathbb{S}_n$ is not solvable.*

PROOF. If $\mathbb{S}_n$ is solvable, then Lemma 1.9 finds $N \triangleleft \mathbb{S}_n$ with N abelian and $|N| > 1$. Let $x \in N$ with $x \neq 1$. Then all of the conjugates of x are in N, and they all commute with each other. The conjugates of x are the elements of $\mathbb{S}_n$ that have the same cycle structure as x.

We will contradict the existence of such an x in cases by the cycle structure of x.

Case 1 If x is a 2-cycle, say $[12]$, then $[12]$ and $[13]$ do not commute.

Case 2 If x is the product of disjoint 2-cycles, and there are at least two 2-cycles, then we can suppose that $x = [12][34] \cdot x'$, where x' is a product of 1-cycles and disjoint 2-cycles none of which involve the points $1, 2, 3, 4$. Form $y = [13][25]y'$, where y' is exactly the same as y, except that 4 replaces 5, either in a 2-cycle or as a fixed point. Then $yxy^{-1}(3) = 5$, so that x, y do not commute.[4]

Case 3 If x involves a k-cycle with $k \geq 3$, then we can suppose that $x = [123\beta]x'$, where β is a sequence of 0 or more points and x' consists of other disjoint cycles, then let $y = [124\delta]y'$, where δ, y' is the same as β, x' except that 3 takes the place of 4, wherever it occurs. Then $yxy^{-1}(2) = 4$, so that x, y do not commute. $\quad\square$

Let's use Eisenstein and a little graphing to get examples. Let E be a splitting field for $X^5 - 10X + 5$ over the rationals. We claim that $\mathrm{Gal}(E/\mathbb{Q})$ is isomorphic to $\mathbb{S}_5$, and so this polynomial is not solvable in radicals.

This polynomial is irreducible by Eisenstein's Criteria. Sketching the graph of $Y = X^5 - 10X + 5$ as in Calculus, we see that it has 3 real roots. Over $\mathbb{R}$ it then factors into 3 linear factors and one quadratic. Because of the quadratic formula, the quadratic splits in $\mathbb{C}$, so now we see that the polynomial splits

---

[4]This is where $n \geq 5$ is crucial. The elements of $\mathbb{S}_4$ written as two disjoint 2-cycles forms an abelian normal subgroup of order 4.

in $\mathbb{C}$. If E is the splitting field in $\mathbb{C}$, then Lemma 8.12 proves that $\mathrm{Gal}(E/\mathbb{Q})$ is isomorphic to $\mathbb{S}_5$. Lemma 8.13 and Lemma 8.10 then tell us that the polynomial is not solvable in radicals.

5. Solutions from Solvable Groups

We want to prove the converse of Lemma 8.10. Throughout this section, let F be a rational field, let $f(X) \in F[X]$, and let K be a splitting field for $f(X)$ over F. We suppose that $\mathrm{Gal}(K/F)$ is a solvable group. Let $n = |K : F|$. We will show that $f(X)$ is solvable in radicals, and the proof can be turned into an algebraic formula for the roots of $f(X)$, at least theoretically.

We will be using properties of solvable groups rather freely, and so a thorough understanding of the proof may require an excursion on your part into group theory.

We need to fatten K a bit.

LEMMA 8.14. *Let L be a splitting field for $X^n - 1$ over K. Then L is Galois over F and $\mathrm{Gal}(L/F)$ is solvable.*

PROOF. We use Theorem 7.5 twice: L is Galois over K, and you have shown that the Galois group is abelian. Furthermore, L is a splitting field for $f(X)(X^n - 1)$ over F, and so L is Galois over F. The Normal Subgroup Theorem then shows that the abelian group $\mathrm{Gal}(L/K)$ is a normal subgroup of $\mathrm{Gal}(L/F)$, and the quotient group is isomorphic to the solvable group $\mathrm{Gal}(K/F)$. Proposition 1.8 shows that $\mathrm{Gal}(L/F)$ is solvable. $\square$

Incorporating L, we assume that $X^n - 1$ splits in K. The main part of our proof shows how the radicals come up; we have heard that Kummer used this argument.

LEMMA 8.15. *Let D be a rational field in which $X^n - 1$ splits. Let E be a Galois extension of D with $|E : D| = p$ where p is a prime divisor of n. Then E is an n-radical extension of D.*

PROOF. If $\alpha \in E$ with $\alpha \notin D$, then the fact that $|E : D|$ is prime forces that $E = D[\alpha]$. The problem with this is that α^p does not have to be in D. We must modify α in such a way as to get a p-th root of an element of D.

We know that if a group has prime order, then it is cyclic, and so

$$\mathrm{Gal}(E/D) = \langle \sigma \rangle \quad \text{for some} \quad \sigma \in \mathrm{Gal}(E/D)$$

Recall the hypothesis that $X^n - 1$ splits in D. You have shown that because D is a rational field, this polynomial has n roots, all of which are powers of some particular root v, and v has multiplicative order n. Since p divides n, the element $z = v^{n/p}$ has multiplicative order p, and so $X^p - 1$ has p roots in D. Here is a trick: let

$$\beta = \sum_{j=0}^{p-1} z^j \sigma^j(\alpha)$$

and then the fact that $z^p = 1$ leads to

(8.2) $$\sigma(\beta) = z^{-1}\beta$$

It follows that

$$\sigma(\beta^p) = z^{-p}\beta^p = \beta^p$$

Because β^p is fixed by σ, it is fixed by every power of σ, and so since σ generates $\mathrm{Gal}(E/D)$ we see that β^p is fixed by the entire Galois group. Since E is a Galois extension of D the Galois Correspondence shows that β^p is in D.

The foregoing argument looks like it proves that β is the element which shows that E is a p-radical extension of D. Indeed, the first paragraph of the proof shows that if $\beta \notin D$, then $D(\beta) = E$. Could it be the case that $\beta \in D$? If so, then the fact that D is fixed by σ shows that $\sigma(\beta) = \beta$. But equation

(8.2) says that $\sigma(\beta) = z^{-1}\beta$. Do you see that $z^{-1} \neq 1$, so that the only way the equations can hold simultaneously is to have $\beta = 0$! In fact, it is not too hard to find an example of the construction of β that *does* yield $\beta = 0$. In order to avoid this unpleasant possibility[5] we need to define β more cleverly in the first place. We define a whole bunch of β's:

$$\beta_j = \sum_{k=0}^{p-1} z^{jk}\sigma^k(\alpha) \quad \text{for} \quad 1 \leq j \leq p-1$$

The fact that $(z^{-j})^p = 1$ yields

(8.3) $$\sigma(\beta_j) = z^{-j}\beta_j$$

which again leads to $\sigma(\beta_j^p) = \beta_j^p$ so that $\beta_j^p \in D$. Again, since $z^{-j} \neq 1$, we have $\beta_j \in D$ implies that $\beta_j = 0$. If all the β_j are 0, equation (8.3) shows that z^j are all roots of the polynomial

$$g(X) = \sum_{k=0}^{p-1} X^k\sigma^k(\alpha)$$

But we already know a polynomial of degree $p-1$ which has all the z^j as roots: the polynomial $\Phi_p(X)$. It must be that $g(X)$ is a multiple of $\Phi_p(X)$, so that the X^0 and X^1 coefficients $g(X)$ are equal: $\alpha = \sigma(\alpha)$. This contradicts the fact that α cannot be fixed by σ, for $\alpha \notin D$.

We have proved that some β_j is not in D. It follows that $E = D[\beta_j]$ and so E is a p-radical extension of D. $\square$

We need yet another technical detail. Recall that $X^n - 1$ splits in L; let J be the splitting field for this polynomial over F in L.

LEMMA 8.16. *Let p be a prime divisor of $|L : J|$. Then p divides n.*

---

[5]There are at least two standard textbooks on abstract algebra that ignore this technicality in their proof of solvability in radicals.

PROOF. We have the equations

$$|L : J| \cdot |J : F| = |L : F| = |L : K| \cdot |K : F| = |L : K| \cdot n$$

We will show that $|L : K|$ divides $|J : F|$ and then it will follow that $|L : J|$ divides n, and this will prove the Lemma.

Notice that both L and J are defined as splitting fields for $X^n - 1$: the field L is a splitting field over K and the field J is a splitting field over F. To show that $|L : K|$ divides $|J : F|$, we will show how $\mathrm{Gal}(L/K)$ may be viewed as a subgroup of $\mathrm{Gal}(J/F)$. Lagrange's Theorem will then show that $|\mathrm{Gal}(L/K)|$ divides $|\mathrm{Gal}(J/F)|$. Because L is Galois over K and J is Galois over F, this will be the same as the statement that $|L : K|$ divides $|J : F|$.

Theorem 5.9 shows that the set of n-th roots of 1 in L constitutes a cyclic group $\langle \beta \rangle$ (where $\beta^n = 1$). Clearly, $L = K[\beta]$ and $J = F[\beta]$. Let $\sigma \in \mathrm{Gal}(L/K)$. Then σ maps β to a power of itself, and it follows that σ maps J to J. Thus σ can be viewed as an element of $\mathrm{Gal}(J/F)$. More formally, we define $\overline{\sigma}$ to be the restriction of σ to J. Then it is easy to see that $\overline{\sigma}$ is a one to one group homomorphism of $\mathrm{Gal}(L/K)$ into $\mathrm{Gal}(J/F)$. This completes the proof. $\square$

We will now show that L is a repeated radical extension of F. Since $f(X)$ splits in K and $K \subseteq L$, this will show that $f(X)$ is solvable in radicals and that it splits in a repeated radical extension of F.

LEMMA 8.17. *The field L is a repeated radical extension of F.*

PROOF. We know that J is an n-radical extension of F. We will show that L is an repeated radical extension of J, and the result will follow.

Since $\mathrm{Gal}(L/J)$ is a normal subgroup of the solvable group $\mathrm{Gal}(L/F)$, Proposition 1.7 shows that $\mathrm{Gal}(L/J)$ is solvable. By the definition of solvable, and by the fact each abelian group has a normal subgroups for each order dividing the order of the abelian group, there are subgroups

$$\mathrm{Gal}(L/J) = N_0 \supseteq N_1 \supseteq \ldots \supseteq N_m = \{1\}$$

where N_j is a normal subgroup of N_{j-1} with $|N_{j-1} : N_j|$ prime, for $1 \le j \le m$. Define fields $L_j = \mathrm{Fix}(N_j)$ and observe that

$$J = L_0 \subseteq L_1 \subseteq \ldots \subseteq L_m = L$$

and L_j is Galois over L_{j-1} for each j.

We claim that each L_j is an n-radical extension of L_{j-1}. Indeed, $J \subseteq L_{j-1}$ shows that $X^n - 1$ splits in L_{j-1}. The prime $p = |L_j : L_{j-1}|$ divides $|L : J|$ and so p divides n by Lemma 8.16. Lemma 8.15 now proves that L_j is a p-radical extension of L_{j-1}. This completes the proof of the lemma. $\qquad\square$

This completes the proof of Galois' Solvability Theorem.

6. The Fundamental Theorem of Algebra

A field F is *algebraically closed* if every polynomial in $F[X]$ splits in F. The Fundamental Theorem of Algebra says that $\mathbb{C}$ is algebraically closed. Throughout the eighteenth century many attempts were made to prove this theorem. Gauss' PhD dissertation, published in 1799, contained an essentially correct proof with a significant gap. Many years later Ostrowski showed how to patch up that argument.[6] Gauss published a different, correct proof in 1816.[7] We will give an algebraic proof commonly attributed to Emil Artin.

THEOREM 8.18. *Let E be a finite extension of $\mathbb{R}$. Then $|E : \mathbb{R}|$ is a power of 2.*

---

[6]A discussion of the gap in Gauss' dissertation and of Ostrowski's work filling the gap is found in Stephen Smale's paper *The Fundamental Theorem of Algebra and Complexity Theory*, in the Bulletin (new series) of the American Mathematical Society, Volume 4, Number 1 (1981).

[7]An English translation of this Latin work appears in [**13**, p.292ff].

PROOF. Since E is a rational field, we can write $E = \mathbb{R}[\alpha]$ for some $\alpha \in E$. Let $f(X)$ be the minimal polynomial for α over $\mathbb{R}$, and let L be a splitting field for $f(X)$ over E. Then L is a splitting field for $f(X)$ over $\mathbb{R}$ and so L is Galois over $\mathbb{R}$. Since $\mathbb{R} \subseteq E \subseteq L$ it is enough to show that $|L : \mathbb{R}|$ is a power of 2. In other words, forget E.

Write $|L : \mathbb{R}| = q \cdot 2^e$ where q is odd. Let $G = \mathrm{Gal}(L/\mathbb{R})$ so that $|G| = q \cdot 2^e$ and by Sylow's Theorem there is a subgroup H in G of order 2^e. Let $F = \mathrm{Fix}(H)$, and then the Galois Correspondence shows that $|F : \mathbb{R}| = q$. By the Primitive Element Theorem we can write $F = \mathbb{R}[\beta]$ for some $\beta \in F$. If $g(X)$ is the minimal polynomial for β over $\mathbb{R}$ then $\deg(g(X)) = q$. Since q is odd, we know that $g(X)$ has a real root.[8] Being irreducible, $g(X)$ must have degree 1. Thus $q = 1$, and so $|L : \mathbb{R}|$ is a power of 2, as needed.   □

Armed with Theorem 8.18, we can prove the Fundamental Theorem. Because polynomials factor into irreducibles, it suffices to show that the irreducible elements of $\mathbb{C}[X]$ have degree 1. Let $f(X)$ be one such irreducible, and let E be a splitting field for $f(X)$ over $\mathbb{C}$. Then E is a Galois extension of $\mathbb{C}$; let $G = \mathrm{Gal}(E/\mathbb{C})$.

Now $|E : \mathbb{R}| = 2|E : \mathbb{C}|$, so that the conclusion of Theorem 8.18 shows that $|E : \mathbb{C}|$ is a power of 2. Assume that it is at least 2, and then Sylow's Theorem gives G a subgroup H of order $|G|/2$, whence $|G : H| = 2$. Put $F = \mathrm{Fix}(H)$, and then the Galois Correspondence shows that $|F : \mathbb{C}| = 2$. Write $F = \mathbb{C}[\alpha]$ and then α is a root of an irreducible polynomial $g(X)$ of degree 2. But the quadratic formula shows that the roots of $g(X)$ are in $\mathbb{C}$, a contradiction. We conclude that $|E : \mathbb{C}| = 1$, and so $f(X)$ splits in $\mathbb{C}[X]$. This completes the proof.

---

[8]This is our one and only use of the (analytic) completeness property of the real numbers in the proof of the Fundamental Theorem.

7. Problems

1. Find the integer coefficients of $\Phi_{24}(X)$.

2. Let $z = \exp(2\pi i/18)$. Find $\alpha \in \mathbb{Q}(z)$ such that $|\mathbb{Q}(\alpha) : \mathbb{Q}| = 3$ and find the minimal polynomial for α over $\mathbb{Q}$. (Hint: The degree of $\Phi_{18}(X)$ is 6. You are looking for a subgroup of the Galois group of order 2.)

3. Suppose that the field F contains at least one root of every non-constant polynomial in $F[X]$. Show that F is algebraically closed.

4. Let Q be the set of elements of $\mathbb{C}$ that are algebraic over $\mathbb{Q}$. Show that Q is an algebraically closed field.

5. Show that $\mathbb{Q}[X]$ is countable, and so the field Q, defined in the previous problem, is countable. Conclude that $\mathbb{R}$ has an element that is not algebraic over $\mathbb{Q}$. Such a number is said to be *transcendental*.

Bibliography

[1] Carl B. Boyer, *The History of the Calculus and its Conceptual Development*, Dover Publications, 1959.

[2] Robert D. Carmichael, *The Theory of Numbers and Diophantine Analysis*, Dover Publications, 1915.

[3] L.E. Dickson, *History of the Theory of Numbers*, Volumes I-III, Chelsea Publishing, 1923.

[4] P.G.L. Dirichlet, *Lectures on Number Theory*, translated by John Stillwell, The American Mathematical Society, 1999.

[5] Harold M. Edwards, *Fermat's Last Theorem: A Genetic Introduction to Algebraic Number Theory*, Springer-Verlag, 1977.

[6] David Eisenbud, *Commutative Algebra with a View Toward Algebraic Geometry*, Springer, 1995.

[7] Euclid, *The Thirteen Books of the Elements*, Volumes 1-3, translated by Sir Thomas L. Heath, Dover Publications, 1956.

[8] Èvariste Galois, *Analyse d'un Mémoire sur la résolution algébrique des équations*, and *Note sur la résolution des équations numériques* and *Sur la théorie des nombres*, Bulletin des Sciences mathématiques, Volume XIII (1830), p.271, 413, 428.

[9] Carl Friedrich Gauss, *Disquisitiones Arithmeticae*, translated by Arthur A. Clarke, Yale University Press, 1965.

[10] Paul R. Halmos, *Naive Set Theory*, D. Van Nostrand, 1960.

[11] Alan Parks, *Foundations of Algebra*, Alas Publishing, 2011.

[12] Alan Parks, *Foundations of Analysis*, Alas Publishing 2012.

[13] David Eugene Smith, *A Source Book in Mathematics*, Dover Publications, 1959.

Index

$C(x)$, centralizer of x in a group, 14

$F[a]$, ring obtained by adjoining a to F, 121

G', derived subgroup of G, 16

G/H, set of cosets in a group, 11

I_n, the $n \times n$ identity matrix, 9

$N \lhd G$, that N is a normal subgroup of G, 11

R/I, quotient ring of ideal I, 34

$R[X]$, polynomials over R, 93

$R \oplus S$, direct sum of rings, 48

U_n, the group of units mod n, 9

$Z(G)$, center of a group G, 14

$\Phi_n(X)$, n-th cyclotomic polynomial, 149

$\binom{n}{k}$, binomial coefficients, 7

$\exp(i \cdot \theta)$, the complex exponential function, 3

$\mathfrak{F}(A, R)$, ring of functions, 50

$\mathfrak{M}(2, R)$, 2×2 matrices over R, 27

$\langle x \rangle$, cyclic group generated by x, 10

$\mathbb{C}$, the complex numbers, 1

$\mathbb{N}$, non-negative integers, 61

$\mathbb{N}$, the non-negative integers, 1

$\mathbb{Q}$, the rational numbers, 1

$\mathbb{R}$, the real numbers, 1

$\mathbb{S}_R$, permutations on R, 10

$\mathbb{Z}$, the integers, 1

$\mathbb{Z}_n$, integers mod n, 2

$\mathrm{Fix}(H)$, fixed field of H, 140

$\mathrm{Gal}(E/F)$, Galois group of E over F, 136

$\ker(f)$, kernel of f, 44

$\overline{z}$, complex conjugate, 5

$\mathrm{End}(R)$, endomorphism ring, 52

$\mathrm{GL}(n, p)$, invertible $n \times n$ matrices mod p., 9

$\mathrm{SL}(n, p)$, $n \times n$ matrices mod p of determinant 1., 10

$\mathrm{cl}(x)$, conjugacy class of x in a group, 14

$\deg(f)$, degree, 94

$|E : F|$, degree of E over F, 117

$|S|$, the order of S, 1

$|z|$, complex number modulus, 3

$a \equiv b \bmod n$, congruence mod n, 2

$n * r$, n copies of r added, 29

abelian, group, 9

adjoining, an algebraic element to a field, 121

algebraic integer, 133

algebraic norm, 72

algebraic, over a field, 120

algebraically closed, field, 174

Ascending Chain Lemma, 63

associates, 56

automorphism, of a field, 136

automorphism, of a ring, 45

basis, 116

binomial coefficients, 7

Binomial Theorem, 8

canonical homomorphism, in a ring, 43